Practical Design of Experiments (DOE)

Also available from ASQ Quality Press:

Practical Attribute and Variable Measurement Systems Analysis (MSA): A Guide for Conducting Gage R&R Studies and Test Method Validations
Mark Allen Durivage

Practical Engineering, Process, and Reliability Statistics
Mark Allen Durivage

The Certified Pharmaceutical GMP Professional Handbook
FDC Division and Mark Allen Durivage, editor

Design of Experiments with Minitab
Paul Mathews

Reliability Data Analysis with Excel and Minitab
Kenneth S. Stephens

The Metrology Handbook, Second Edition
Jay L. Bucher, editor

The Certified Quality Engineer Handbook, Third Edition
Connie M. Borror, editor

The Certified Six Sigma Green Belt Handbook, Second Edition
Roderick A. Munro, Govindarajan Ramu, and Daniel J. Zrymiak

The Certified Six Sigma Black Belt Handbook, Second Edition
T. M. Kubiak and Donald W. Benbow

The Certified Reliability Engineer Handbook, Second Edition
Donald W. Benbow and Hugh W. Broome

The Certified Quality Inspector Handbook, Second Edition
H. Fred Walker, Ahmad K. Elshennawy, Bhisham C. Gupta, and Mary McShane Vaughn

The Certified Quality Technician Handbook, Second Edition
H. Fred Walker, Donald W. Benbow, and Ahmad K. Elshennawy

Failure Mode and Effect Analysis: FMEA from Theory to Execution, Second Edition
D. H. Stamatis

To request a complimentary catalog of ASQ Quality Press publications,
call 800-248-1946, or visit our website at http://www.asq.org/quality-press.

Practical Design of Experiments (DOE)

A Guide for Optimizing Designs and Processes

Mark Allen Durivage

ASQ Quality Press
Milwaukee, Wisconsin

American Society for Quality, Quality Press, Milwaukee 53203
 Published 2016
Printed in the United States of America
22 21 20 19 18 17 16 5 4 3 2 1

Library of Congress Cataloging-in-Publication Data

Names: Durivage, Mark Allen.
Title: Practical design of experiments (DOE) : a guide for optimizing designs and processes / Mark Allen Durivage.
Description: Milwaukee, Wisconsin : ASQ Quality Press, 2016. | Includes bibliographical references and index.
Identifiers: LCCN 2015049294 | ISBN 9780873899246 (hard cover : alk. paper)
Subjects: LCSH: Engineering—Statistical methods. | Acceptance sampling. | Quality control—Statistical methods. | Distribution (Probability theory) | Science—Methodology.
Classification: LCC TA340 .D868 2016 | DDC 001.4/34—dc23
LC record available at http://lccn.loc.gov/2015049294

ISBN: 978-0-87389-924-6

ASQ Mission: The American Society for Quality advances individual, organizational, and community excellence worldwide through learning, quality improvement, and knowledge exchange.

Quality Press
600 N. Plankinton Ave.
Milwaukee, WI 53203-2914
Email: books@asq.org
ASQ Excellence Through Quality™

Table of Contents

List of Figures and Tables

Preface

This book—a result of 30 years of quality-related work experience—was written to aid quality technicians and engineers. To that end, the intent of this book is to provide the quality professional working in virtually any industry a quick, convenient, and comprehensive guide to properly conducting *design of experiments* (DOE) for the purpose of process optimization.

This book is intended for people who have never been exposed to design of experiments, been intimidated in their attempts to learn about DOE, or have not appreciated the potential of this family of tools in their process improvement and optimization efforts. This is a practical introduction to the basics, and is not intended to provide complete coverage of DOE. When one becomes familiar with and begins applying the basic principles presented in the book, they should be easily encouraged to go on to more-advanced principles in DOE. Once successful with DOE, one will rarely need prodding to continue to learn more about this powerful tool. Every effort has been made to simplify the approach and minimize the complexity of the material. This book will use a simple statistical calculator rather than sophisticated software or spreadsheets. The book also assumes a basic knowledge of statistical techniques and process control. Inevitably, there are some points that might have been covered with more statistical depth. It is the author's strong recommendation that anyone who completes this book immediately continue on to a more in-depth reference (some excellent ones are identified in the Bibliography of this book), both to learn about areas that are not covered here and to broaden the reader's depth of knowledge of DOE in general.

This book will be a useful reference when preparing for and taking many of the ASQ quality certification examinations, including the Certified Quality Technician (CQT), Certified Six Sigma Green Belt (CSSGB), Certified Quality Engineer (CQE), Certified Six Sigma Black Belt (CSSBB), and Certified Reliability Engineer (CRE).

Acknowledgments

I would like to acknowledge the previous work of Larry B. Barrentine in *An Introduction to Design of Experiments: A Simplified Approach*. This book is an expansion of his efforts in an attempt to continue Barrentine's method of presenting DOE studies in a simple, easy-to-follow style. Several sections of this book come directly from his previous work. I have made some changes to clarify and augment some of his points and present the topics in a consistent manner.

I would like to thank those who have inspired, taught, and trained me throughout my academic and professional career. I also wish to recognize my friend and colleague, Scott Kochendoerfer, CQE, for lending his expertise in reviewing this book for accuracy and content. I would also like to express my sincere gratitude to James McLinn, Reliability Consultant at Ops A La Carte, and Stefan Mozar, Director and Adjunct Professor at Guangdong University of Technology, for reviewing the book and providing valuable feedback. Additionally, I would like to thank ASQ Quality Press, especially Matt Meinholz, Acquisitions Editor, and Paul Daniel O'Mara, Managing Editor, for their expertise and technical competence, which made this project a reality. Lastly, I would like to acknowledge the patience of my wife Dawn and my sons Jack and Sam, which allowed me time to research and write *Practical Design of Experiments (DOE): A Guide for Optimizing Designs and Processes.*

Limit of Liability/Disclaimer of Warranty

The author has put forth his best efforts in compiling the content of this book; however, no warranty with respect to the material's accuracy or completeness is made. Additionally, no warranty is made in regard to applying the recommendations made in this book to any business structure or environments. Businesses should consult regulatory, quality, and/or legal professionals prior to deciding on the appropriateness of advice and recommendations made within this book. The author shall not be held liable for loss of profit or other commercial damages resulting from the employment of recommendations made within this book, including special, incidental, consequential, or other damages.

1
Introduction

The English statistician Sir Ronald A. Fisher pioneered the development of design of experiments in the 1920s and 1930s, applying statistical techniques in the study of agriculture. During the 1940s, Robin L. Plackett and J. P. Burman introduced the idea of using smaller, more economical designs for experimentation (fractional factorials). The 1950s saw the introduction of *response surface methodology* (RSM), used in industrial experimentation by George E. P. Box and K. B. Wilson. During the same period of time, Genichi Taguchi introduced methods for improving the quality of manufactured goods, applying the loss function and signal-to-noise ratios to experimentation.

The purpose of this introduction to the design of experiments (DOE) is to showcase the power and utility of this statistical tool while explaining how to plan and analyze an experiment. It is also an attempt to dispel the conception that DOE is reserved only for those with advanced mathematics training. It will be demonstrated that DOE is primarily a logic tool that can be easily grasped and applied, requiring only basic math skills. While software or spreadsheets would make the calculations more painless and provide greater versatility, it is necessary to understand what the software is doing. To this end, software is not used with this text; an inexpensive scientific calculator is used instead to insure that the basics are learned. This is by no means a complete study of the broad field of DOE. The intent is to introduce the basics, persuade the reader of the power of this tool, and then recommend resources for further study. The material covered will still be sufficient to support a large percentage of the experiments one may wish to perform.

The prerequisites of this book are familiarity with the basic concepts of statistics, process stability, statistical process control (SPC), and measurement systems analysis (MSA). As in any process improvement activity, it is necessary to recognize that a process is made up of input variables, process variables, and output measures (see Figure 1.1). The intent is always to improve the output measure, which is labeled as the *response*. There is no direct control on the response variable; in the classical cause-and-effect approach, it is the *effect*.

The causes are what dictate the response (dependent variable, output, *Y*'s). To control the response, one must control the causes (independent variable, inputs, *X*'s), which may be input variables and/or process variables involving the six elements shown in Figure 1.1. (These variables or causes will later be referred to as *factors*.)

For example, there is no control setting in a sales process that allows one to set a sales level. To control sales, one must address those variables that cause sales to change,

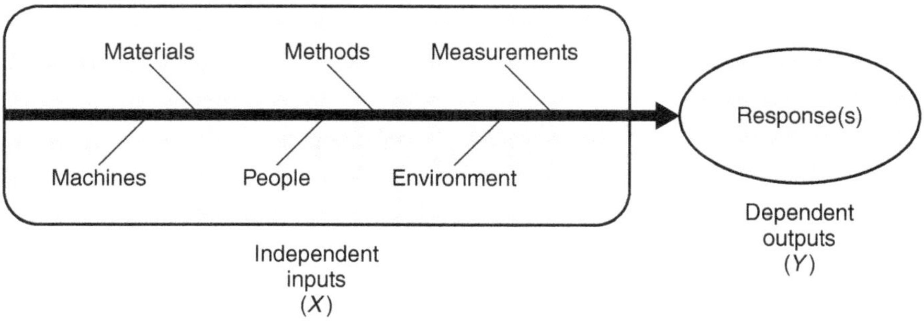

Figure 1.1 Cause-and-effect diagram depicting inputs (*X*'s) and outputs (*Y*'s).

for example, promotional literature, call frequency, pricing policies, credit policies, or personal sales techniques. A process may be very simple, or it may be a complex group of processes.

In concert with this cause-and-effect, or *systems*, approach to the process, the concepts of *process variation* must be understood. Every response demonstrates variation. This variation results from (a) variation in the known input or process variables, (b) variation in the unknown process variables, and/or (c) variation in the measurement of the response variable. The combination of these sources results in the variation of that response. This variation is categorized by the classic SPC tools into two categories: (a) special cause variation—unusual responses compared to previous history; and (b) inherent variation—variation that has been demonstrated as typical of the process.

A side note is needed here on terminology. *Inherent*, or *typical*, *variation* has a variety of labels that are often used interchangeably. In control charting, it is referred to as *common cause variation*. In control systems, it is called *process noise*. In DOE, it is called *experimental error* or *random variation*. To minimize confusion, it will be referred to in this text as either *inherent variation* or *experimental error*.

Control charts are used to identify special cause variation and, hopefully, to identify the process variables or causes that led to such unusual responses. The presence of special causes within an experiment will create problems in reaching accurate conclusions. For this reason, DOE is more easily performed after the process has been stabilized using SPC tools. The presence of inherent variation also makes it difficult to draw conclusions. (In fact, that is one of the definitions of statistics: decision making in the presence of uncertainty or inherent variation.) If a process variable causes changes in the response that exceed the inherent variation, we state that the change is *significant*.

Inherent variation can also be analyzed to determine whether the process will consistently meet a specification. The calculation of process capability is a comparison of the spread of the process with the specifications, resulting in test statistics such as C_p and C_{pk}. Figure 1.2 illustrates the comparison of a process with its upper and lower specification limits.

DOE is the simultaneous study of several process variables. By combining several variables in one study instead of creating a separate study for each, the amount of testing

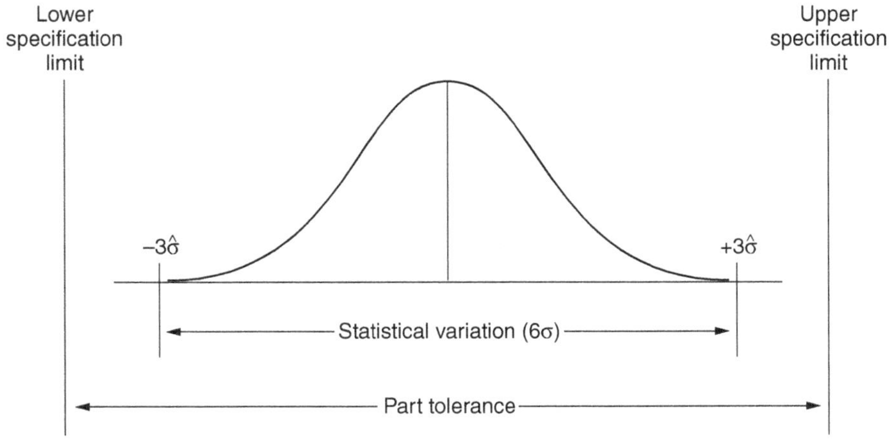

Figure 1.2 Relationship between statistical control limits and product specifications.
Source: M. A. Durivage, *Practical Engineering, Process, and Reliability, Statistics*, Milwaukee: ASQ Quality Press, 2014. Used with permission.

required will be drastically reduced, and greater process understanding will result. This is in direct contrast to the typical *one-factor-at-a-time* (OFAT) approach, which limits understanding and wastes data. Additionally, OFAT studies cannot be assured of detecting the unique effects of combinations of factors (a condition later to be defined as an *interaction*).

DOE includes the entire scope of experimentation, including defining the output measure(s) that one desires to improve, the candidate process variables that one will change, procedures for experimentation, actual performance of the experiment, and analysis and interpretation of the results. The objectives of the experimenter in a DOE are to learn how to:

- Maximize the response
- Minimize the response
- Adjust the response to a nominal value
- Reduce process variation
- Make the process robust (that is, make the response insensitive to uncontrollable changes in the process variables)
- Determine which variables are important to control and which are not

The *basic experimental procedure* is a series of basic logical steps that must be addressed as one prepares to launch a DOE:

1. What is the process to be studied? How broadly or narrowly is it defined? A flowchart is a good tool for this analysis.

2. What is the response? What needs to be improved? Should there be more than one response? Note: Additional responses are free, costing only the measurements!
3. What is the measurement precision? Is there bias in the measurement system? Has a measurement systems analysis been completed? Is it adequate?
4. Generate candidate factors. This is best done with a small team using brainstorming after a review of all available data and information on the process and response variables. A cause-and-effect diagram and a flowchart of the process are useful tools to use while brainstorming. The trick is to be innovative, to think outside usual boundaries, and yet not try to reinvent proven technology. Provide opportunities for surprises! The team should be knowledgeable about the issues and follow the rules for brainstorming.
5. Determine the levels for the factors selected for the DOE. In screening experiments, the rule is to have levels broadly spaced but not to the point of being foolhardy. In refining experiments, levels will be much tighter and will require more replication.
6. Select the experimental design. This is the set of treatments or runs that will be performed. This also includes deciding on the amount of replication. Finally, the randomized order of the trials is determined. (Randomization is the insurance policy against misleading conclusions due to outside influence during the experiment.)
7. Establish a plan to control (or at least monitor) extraneous variables.
8. Perform the experiment according to the design. The DOE must be carried out per its design. Identify trial materials carefully. Keep good notes.
9. Analyze, draw conclusions, and assess process impact. What process variables can be changed—and how—to improve the process?
10. Verify and document the new process as defined by the experiment.
11. Propose the next study for continuation of this project, or declare the project complete. Make sure that all reports that go beyond the team are in language and terminology that are easily understood.

Note: It is extremely important that prior to and after performing DOE a line clearance is executed to prevent mix-ups and/or comingling of products, packaging, and labeling.

The *nine steps for analysis of effects* are shown in Figure 1.3. This will be the basic flow for all experiments presented in this book. There will be times when some of the steps cannot be completed. For instance, steps 3, 4, 5, and 6 are not used when conducting experiments without repetitions, replicates, attribute data, ordered categorical data, and Taguchi's signal-to-noise (S/N) ratios. In these cases, a half-normal plot can be used to determine the significant effects. Some of the examples in the book that use steps 3, 4,

1. Calculate the absolute values of the effects
2. Create a Pareto chart of the absolute values of the effects
3. Calculate the standard deviation of the experiment, s_e
4. Calculate the standard deviation of the effects, s_{Eff}
5. Determine the *t*-statistic
6. Calculate the decision limits
7. Determine the significant effects
8. Graph the main effects and any significant interactions
9. Model the significant effects, significant interactions, and individual effects from significant interactions

Figure 1.3 Nine steps for analysis of effects.

5, and 6 will also have an associated half-normal plot for illustrative purposes. It should be noted that the use of statistical decision limits is the preferred method.

For ease of instruction, a review of some the basic statistical tools and techniques is presented, followed by small experiments, which are followed by large experiments. In the real world, one would prefer to start with large experiments and progress to smaller ones in order to identify variables that affect the response variables. Terms and definitions are covered as they arise. The terminology used in DOE is often different from the equivalent terms in SPC, and is presented to assure easier readability. The initial example is used to define most of the unique terminology and many of the analytical techniques. It is suggested that the reader review the Glossary following the Appendixes.

Disclaimer: All examples in this book are fictitious, and therefore the results should not be used to make product or process decisions.

2

Statistical Tools and Techniques

This chapter will provide an overview of some additional statistical tools and techniques that are utilized when performing DOE.

2.1 DEAN AND DIXON OUTLIER TEST

The *Dean and Dixon outlier test* is a valid method for detecting outliers when the data are normally distributed. To use this test, the data must be ordered from smallest value to largest value. The formulas are dependent on the sample size:

N	**Smallest value test**	**Largest value test**
3 to 7	$r_{10} = \dfrac{X_2 - X_1}{X_n - X_1}$	$r_{10} = \dfrac{X_n - X_{n-1}}{X_n - X_1}$
8 to 10	$r_{11} = \dfrac{X_2 - X_1}{X_{n-1} - X_1}$	$r_{11} = \dfrac{X_n - X_{n-1}}{X_n - X_2}$
11 to 13	$r_{21} = \dfrac{X_3 - X_1}{X_{n-1} - X_1}$	$r_{21} = \dfrac{X_n - X_{n-2}}{X_n - X_2}$
14 to 30	$r_{22} = \dfrac{X_3 - X_1}{X_{n-2} - X_1}$	$r_{22} = \dfrac{X_n - X_{n-2}}{X_n - X_3}$

where

n = Number of data points

(See Appendix A, Critical Values of the Dean and Dixon Outlier Test.)

Example: Given the following data points, verify whether the smallest and largest values are outliers at 95% confidence:

1, 3, 6, 7, 8, 9, 10, 11, 12, 23

Since there are 10 data points (n = 10) we must use the second set of equations for 8 to 10 to detect the presence of an outlier.

To detect whether the smallest value is an outlier:

$$r_{11} = \frac{X_2 - X_1}{X_{n-1} - X_1} = \frac{3-1}{12-1} = 0.182 \quad (r_{11\ \alpha=0.05} \text{ table value is } 0.477)$$

Since r_{11} calc is < r_{11} crit, we can conclude that the smallest value is not an outlier at the 95% confidence level.

To detect whether the largest value is an outlier:

$$r_{11} = \frac{X_n - X_{n-1}}{X_n - X_2} = \frac{23-12}{23-3} = 0.550 \quad (r_{11\ \alpha=0.05} \text{ table value is } 0.477)$$

Since r_{11} calc is > r_{11} crit, we can conclude that the largest value is an outlier at the 95% confidence level.

2.2 HYPOTHESIS TESTING

Type I and Type II Errors

When testing hypotheses, we usually work with small samples from large populations. Because of the uncertainties of dealing with sample statistics, decision errors are possible. There are two types of errors: type I errors and type II errors.

A *type I error* occurs when the null hypothesis is rejected when it is, in fact, true. In most tests of hypotheses, the risk of committing a type I error (known as the *alpha* [α] *risk*) is stated in advance along with the sample size to be used (see Table 2.1).

A *type II error* occurs when the null hypothesis is accepted when it is false. If the means of two processes are, in fact, different, and we accept the hypothesis that they are equal, we have committed a type II error. The risk associated with type II errors is referred to as the *beta (β) risk.* This risk is different for every alternative hypothesis and is usually unknown except for specific values of the alternative hypothesis (see Table 2.1).

Alpha (α) and Beta (β) Risks

There is an associated risk with each of the errors just described.

The risk of rejecting the null hypothesis when it is true is the *alpha (α) risk.* The experimenter usually determines the risk before running the test. Common levels of alpha (α) risk are 0.10, 0.05, and 0.01.

An alpha (α) risk of 0.10 means that, in the long run, the null hypothesis will be rejected when it is true (type I error) 10 times out of a hundred; an alpha (α) risk of 0.05 means that, in the long run, the null hypothesis will be rejected when it is true (type I

Table 2.1 Hypothesis truth table.

Decision on null hypothesis	Null hypothesis is true	Null hypothesis is false
Accept (Fail to reject)	Correct decision Probability = $1 - \alpha$	Type II error Probability = β
Reject	Type I error Probability = α	Correct decision Probability = $1 - \beta$

error) five times out of a hundred; and an alpha (α) risk of 0.01 means that, in the long run, the null hypothesis will be rejected when it is true (type I error) one time out of a hundred, and so on.

Type I and type II errors cannot occur at the same time. Type I error can happen only if H_0 is true, and type II error can happen only when H_0 is false. If the probability of type I error is high, then the probability of type II error is low.

The statistical power of a test is also another consideration when conducting tests of hypotheses. The power of a test is $1 - \beta$, and is expressed as a probability ranging from 0 to 1. The higher the power, the more sensitive the test is in detecting differences where differences exist between the hypothesized parameter and its true value. Increasing the sample size, reducing the variance, increasing the effect size index, or decreasing the alpha level (α) will increase the statistical power.

Apportionment of Risk in Hypothesis Testing

Tests of hypotheses and the apportionment of the risk usually fall into two categories: one-tail tests, where all of the risk is in a single tail of the distribution, and two-tail tests, where the risk is evenly split between the two tails of the distribution. When we conduct tests of hypotheses, we must calculate a value and compare it to a critical value from a reference table. It is possible we can fail to reject the double-tailed test, the upper-tailed test, and the lower-tailed test null hypotheses.

The Hypothesis Test for a One-Tail (Upper-Tailed) Test

The hypothesis for a one-tail (upper-tailed) test might be stated as

$$H_0 : \mu_1 \le \mu_2 \qquad H_0 : \sigma_1^2 \le \sigma_2^2$$
$$\text{or}$$
$$H_A : \mu_1 > \mu_2 \qquad H_A : \sigma_1^2 > \sigma_2^2$$

The null hypothesis (H_0) is that $\mu_1 \le \mu_2$, or $H_0 : \sigma_1^2 \le \sigma_2^2$, while the alternative hypothesis (H_A) is that $\mu_1 > \mu_2$, or $H_A : \sigma_1^2 > \sigma_2^2$.

We reject the null hypothesis (H_0) if $Z_{Calculated} > Z_{Critical}$.

The hypothesis for a one-tail (upper-tailed) test, as depicted in Figure 2.1, is used when one suspects that the alternative hypothesis (H_A) is greater than the null hypothesis (H_0).

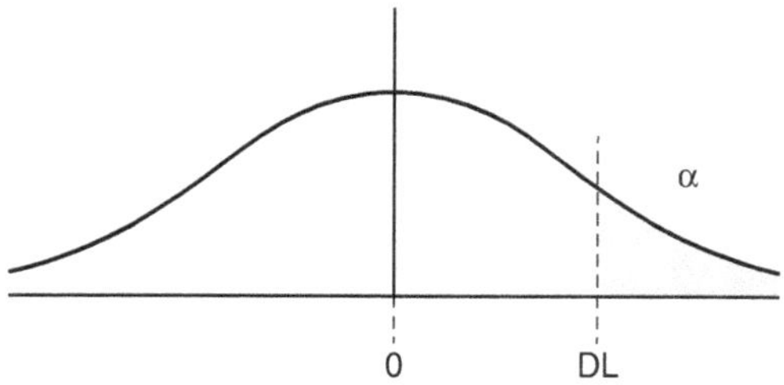

Figure 2.1 Representation of a one-tail (upper-tailed) test.

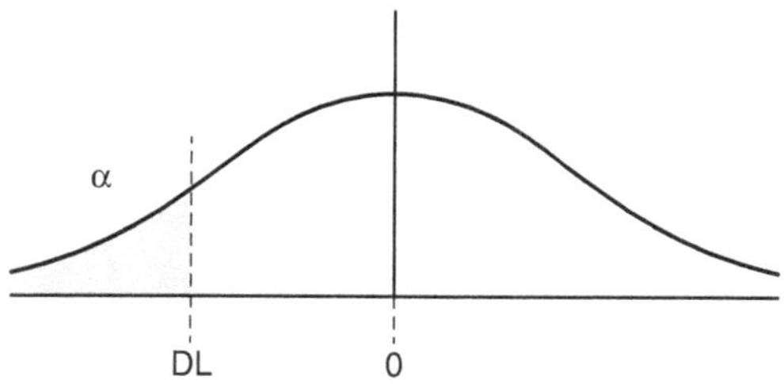

Figure 2.2 Representation of a one-tail (lower-tailed) test.

The Hypothesis Test for a One-Tail (Lower-Tailed) Test

The hypothesis for a one-tail (lower-tailed) test might be stated as

$$\begin{matrix} H_0 : \mu_1 \geq \mu_2 \\ H_A : \mu_1 < \mu_2 \end{matrix} \quad \text{or} \quad \begin{matrix} H_0 : \sigma_1^2 \geq \sigma_2^2 \\ H_A : \sigma_1^2 < \sigma_2^2 \end{matrix}$$

The null hypothesis (H_0) is that $\mu_1 \geq \mu_2$, or $H_0 : \sigma_1^2 \geq \sigma_2^2$, while the alternative hypothesis (H_A) is that $\mu_1 < \mu_2$, or $H_A : \sigma_1^2 < \sigma_2^2$.

We reject the null hypothesis (H_0) if $Z_{Calculated} < Z_{Critical}$.

The hypothesis for a one-tail (lower-tailed) test, as depicted in Figure 2.2, is used when one suspects that the alternative hypothesis (H_A) is less than the null hypothesis (H_0).

The Hypothesis Test for a Two-Tail Test

The hypothesis for a two-tail test might be stated as

$$\begin{matrix} H_0 : \mu_1 = \mu_2 \\ H_A : \mu_1 \neq \mu_2 \end{matrix} \quad \text{or} \quad \begin{matrix} H_0 : \sigma_1^2 = \sigma_2^2 \\ H_A : \sigma_1^2 \neq \sigma_2^2 \end{matrix}$$

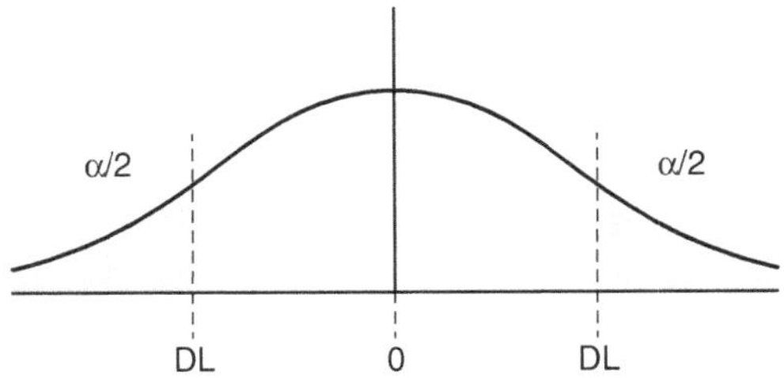

Figure 2.3 Representation of a two-tail test.

The null hypothesis (H_0) is that $\mu_1 = \mu_2$, or $H_0 : \sigma_1^2 = \sigma_2^2$, while the alternative hypothesis (H_A) is that $\mu_1 \neq \mu_2$, or $H_A : \sigma_1^2 \neq \sigma_2^2$.

We reject the null hypothesis (H_0) if $|Z_{\text{Calculated}}| > \pm Z_{\text{Critical}}$.

The hypothesis for a two-tail (double-tailed) test, as depicted in Figure 2.3, is used when one suspects that the alternative hypothesis (H_A) is not equal to the null hypothesis (H_0).

The Hypothesis Test Conclusion Statements

Once a hypothesis test has been conducted, it is necessary to draw a conclusion and formulate a summary statement of the findings.

Example 1: *(Fail to reject the null hypothesis).* Since Z calculated is less than Z critical at the 0.05 α (alpha) level, there is insufficient evidence to reject the null hypothesis that $H_0 : \mu_1 = \mu_2$.

Example 2: *(Reject the null hypothesis).* Since Z calculated is greater than Z critical at the 0.05 α (alpha) level, there is sufficient evidence to reject the null hypothesis $H_0 : \sigma_1 = \sigma_2$ in favor of the alternative hypothesis that $H_A : \sigma_1^2 \neq \sigma_2^2$.

Testing for a Difference between Two Observed Variances Using Sample Data

When comparing the variances of two populations using sample data, the F-test is appropriate if both populations are normally distributed. While the t-test is less sensitive to small departures from normality, these can significantly affect the risks involved with the F-test. Another assumption of this test is that the samples are independent. The formula for the F-test is

$$F = \frac{s_1^2}{s_2^2}$$

where

$s_1^2 > s_2^2$

$\text{df}_1 = n_1 - 1$

$\text{df}_2 = n_2 - 1$

s_1 = Standard deviation of population 1

s_2 = Standard deviation of population 2

n_1 = Number of samples drawn from population 1

n_2 = Number of samples drawn from population 2

$F\alpha$, $\text{df}_1 = n_1 - 1$, $\text{df}_2 = n_2 - 1$ = values of the F-distribution. (See Appendix B, Percentages of the F-Distribution)

The larger variance is always placed in the numerator. This results in a one-tail test that simplifies both calculation and interpretation.

Example: Two competing machine tool manufacturers presented equipment proposals to a customer. After reviewing both proposals, a test was conducted to determine whether the variances of the product produced by the two machines are equivalent. A sample of 25 from machine 1 produced a variance of 0.1211, and 30 units from machine 2 resulted in a variance of 0.0707. At the 95% confidence level, test the hypothesis that the variances are equal.

$H_0 : \sigma_1 = \sigma_2$ (null hypothesis two-tail test)

$H_A : \sigma_1 \neq \sigma_2$ (alternative hypothesis two-tail test)

$F\alpha$, $\text{df}_1 = n_1 - 1$, $\text{df}_2 = n_2 - 1 = 1.90$

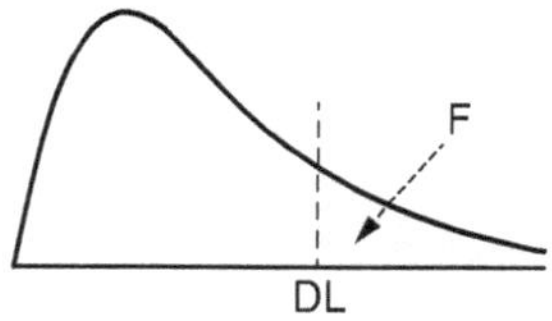

***F* calculated**		***F* critical**
$F = \frac{s_1^2}{s_2^2} = \frac{0.1211}{0.0701} = 1.728$	$<$	1.90

Since F calculated is less than F critical at the 0.05 α (alpha) level, there is insufficient evidence to reject the null hypothesis that $H_0 : \sigma_1 = \sigma_2$.

2.3 NORMAL PROBABILITY PLOTS

Normal probability plots can be constructed to look for linearity when using one variable. Normal probability plots provide a visual way to determine if a distribution is approximately normal.

If the distribution is close to normal, the plotted points will lie close to a line. Systematic deviations from a line indicate a nonnormal distribution:

- *Right skew.* If the plotted points appear to bend up and to the left of the normal line, this indicates a long tail to the right (Figure 2.4).
- *Left skew.* If the plotted points bend down and to the right of the normal line, this indicates a long tail to the left (Figure 2.5).
- *Short tails.* An S-shaped curve indicates shorter-than-normal tails, that is, less variance than expected (Figure 2.6).

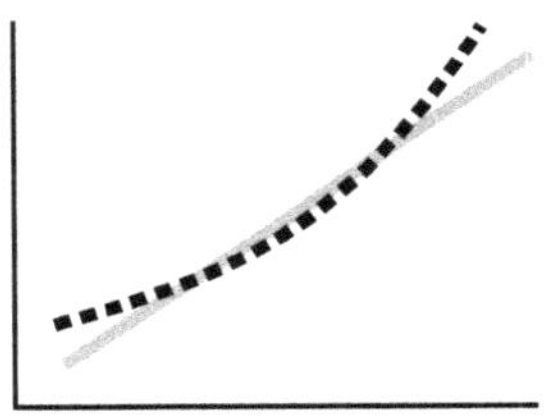

Figure 2.4 Right-skewed distribution.

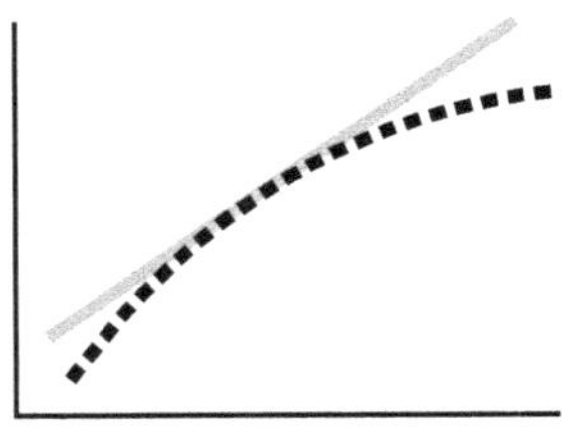

Figure 2.5 Left-skewed distribution.

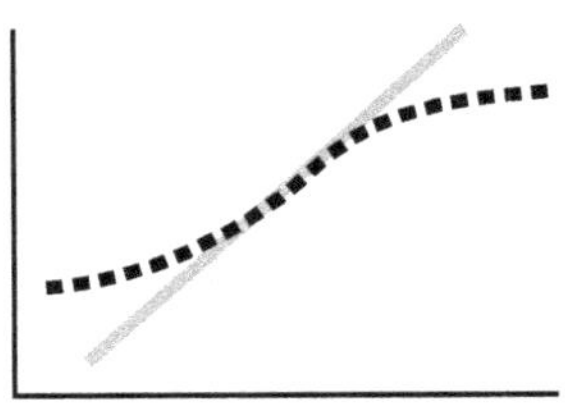

Figure 2.6 Short-tailed distribution.

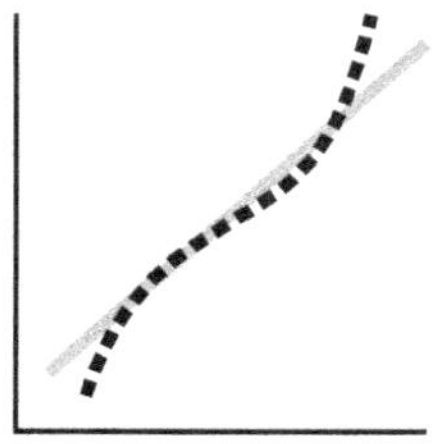

Figure 2.7 Long-tailed distribution.

- *Long tails.* A curve that starts below the normal line, bends to follow it, and ends above it indicates long tails. That is, you are seeing more variance than you would expect in a normal distribution (Figure 2.7).

Normal probability plots are constructed by doing the following:

1. The data are arranged from smallest to largest.
2. The percentile of each data value is determined.
3. From these percentiles, normal calculations are done to determine their corresponding z-scores.
4. Each z-score is plotted against its corresponding data value.

To calculate the percentile, use the following formula:

$$p = \frac{(i - 0.5)}{n}$$

where

i = The individual value

n = The number of data points

Each p-value must be converted to its corresponding z-score (see Table 2.2) using Appendix F, Normal Distribution Probability Points—Area below Z, for p-values ≥ 0.500 and Appendix G, Normal Distribution Probability Points—Area above Z, for p-values $\leq$ 0.500. For example, p 0.07 using Appendix G, $Z = -1.47$, p 0.93 using Appendix F, $z =$ 1.47. In these cases there was not an exact z value that corresponds to the p-value. So one must choose a z- value closest to the p-value. For any p-value below 0.50, the z-value will be negative, and for any p-value above 0.50, the z-value will be positive.

A graphical review of the normal probability plot for strength (Figure 2.8) does not yield a normal distribution for the data points analyzed as they do not generally fall on the line of best fit. As a general rule, if a pencil can be placed on the points and the points are covered, the points can be considered to be from a normal distribution. As in this case, a pencil would not cover the points, and therefore this set of points can be

Table 2.2 Calculation summary.

Strength	*i*	*p*	*z*
41	1	0.07	–1.47
41	2	0.21	–0.79
41	3	0.36	–0.37
43	4	0.50	0.00
43	5	0.64	0.37
44	6	0.79	0.79
50	7	0.93	1.47

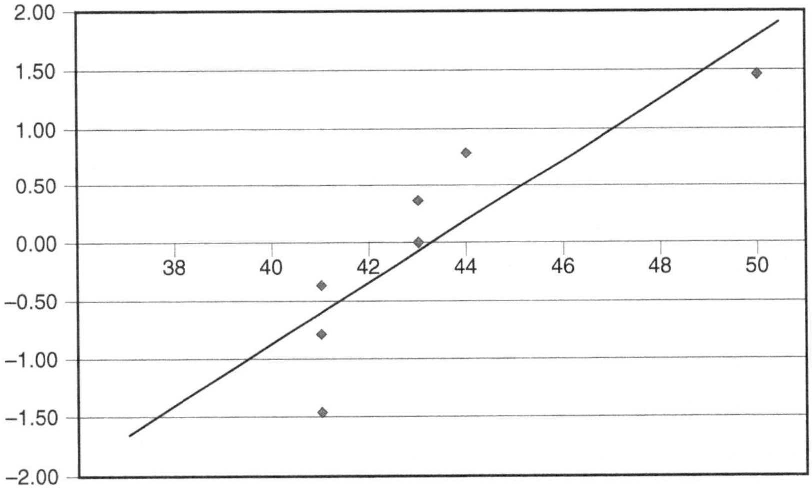

Figure 2.8 Normal probability plot for strength.

described as not coming from a normal distribution. This classic analogy is called the "pencil test" of normality.

2.4 HALF-NORMAL PROBABILITY PLOTS

A half-normal probability plot is a graphical tool that uses the ordered absolute estimated effects to help assess which factors are important and which are unimportant. The effects that are not significant will have plotted points that lie close to a line that has its origin at 0,0 and passes through the cumulative percentage point near 50%. Student's *t*-test is preferred over this method. However, when the experiment does not have replicates, this is the best method to determine the significant effects. When the number of effects is small, as in a two-factor experiment, the results may be erratic using this method.

Half-normal probability plots are constructed by doing the following:

1. The data are arranged from smallest to largest absolute values (see Table 2.3).
2. The cumulative percentage of each data value is determined.
3. Each cumulative percentage is plotted against its corresponding absolute effect value.
4. Construct the line by plotting a line from 0,0 that passes through the cumulative percentage point at 50%. Double the effect value at the 50% cumulative percentage point, which will extend the line to 1.0 or 100%. If this does not appear to be the line of best fit (pencil test), adjust the doubled value up or down as necessary. If there is an even number of effects, interpolate between the two middle-most points.

To calculate the percentile, use the following formula:

$$p = \frac{(i-5)}{n}$$

where

i = The individual value

n = The number of data points

For this example, the effect at 50% is 2.27, which doubled is 5.34. The line is plotted (0,0) and (5.34,1), but is probably not the line of best fit, so another line, represented as a dashed line, is plotted (0,0) and (5.60,1). This dashed line is a much better fit. A graphical review of the half-normal probability plot of effects (Figure 2.9) indicates the 13.84 effect is significant; the data point analyzed does not generally fall on the line of best fit.

Note: A table of cumulative percentage points is included in Appendix D, Cumulative Percentage Points.

Table 2.3 Calculation summary.

Effect	*i*	*p*
0.27	1	0.07
1.29	2	0.21
1.67	3	0.36
2.27	4	0.50
3.91	5	0.64
4.81	6	0.79
13.84	7	0.93

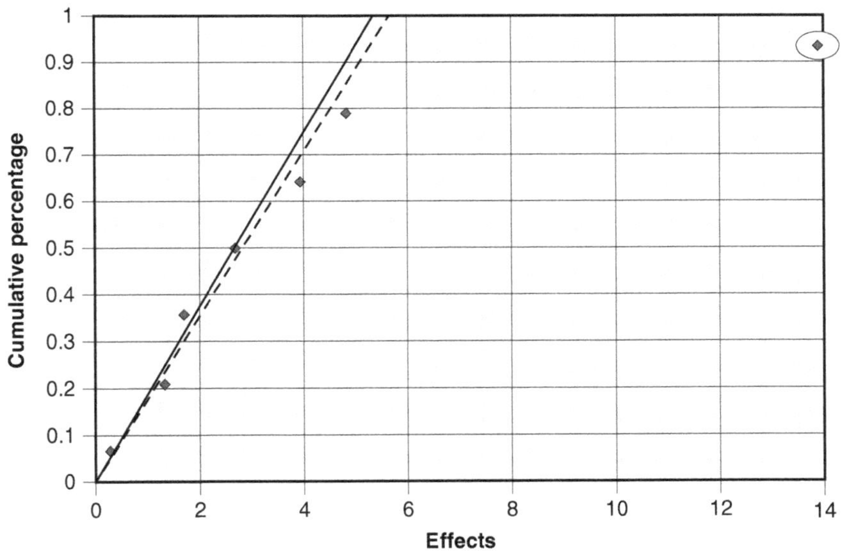

Figure 2.9 Half-normal probability plot for effects.

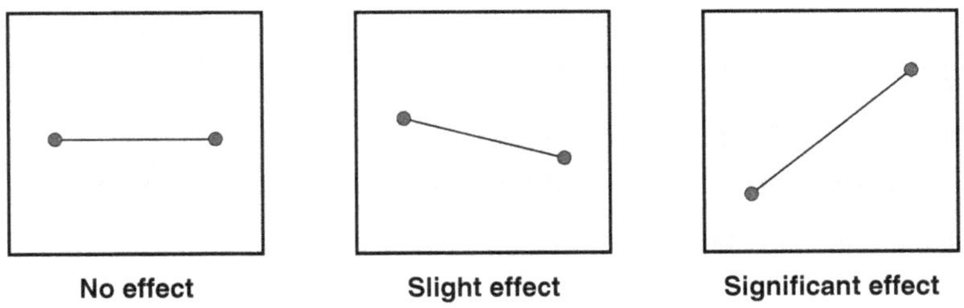

Figure 2.10 Factor effect plots.

2.5 INTERPRETING EFFECT AND INTERACTION PLOTS

Effects are plotted and used to determine how the levels of the factors influence the effect. Interaction plots plot two or more factors to determine whether interactions between the factors are present. There are few simple rules that dictate their interpretation.

Factors effect plot interpretation (see Figure 2.10):

1. If an effect is horizontal, the effect is not significant.
2. If an effect is not horizontal, the effect has more significance as the slope of the line increases. Please note that the direction of the slope of the line does not matter.

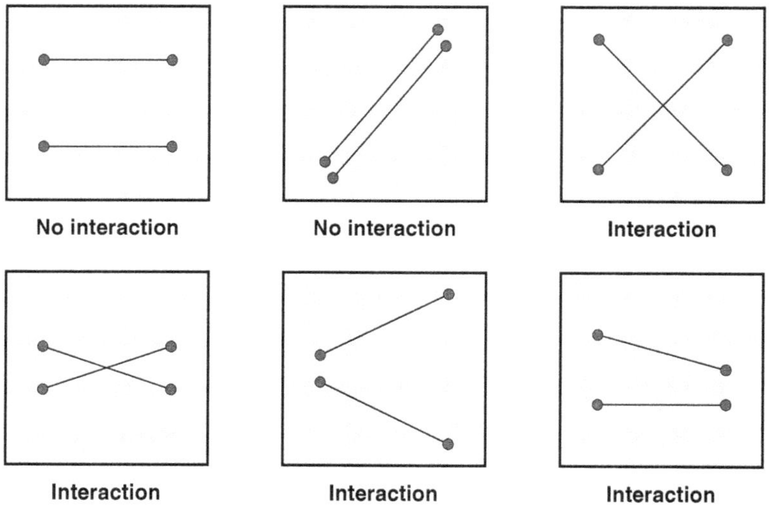

Figure 2.11 Factor interaction plots.

Factor interaction plot interpretation (see Figure 2.11):

1. If the lines are parallel, an interaction is not present, even if the lines are not horizontal. The distance between the lines does not matter.
2. If the lines intersect, or can intersect, an interaction between the factors is present. The slope of the lines and their direction do not matter.
3. A factor with a higher slope is more significant than a factor that is horizontal.

3
ANOVA

Analysis of variance (ANOVA) is a collection of statistical models used to analyze the differences between three or more groups of means and their associated variation among and between groups. In the ANOVA setting, the observed variance in a particular variable is partitioned into components attributable to different sources of variation. In its simplest form, ANOVA provides a statistical test of whether or not the means of three or more groups are equal, and therefore generalizes the t-test to more than two groups. Doing multiple two-sample t-tests would result in an increased chance of committing a type I error. For this reason, ANOVAs are useful in comparing three or more means for statistical significance (http://en.wikipedia.org/wiki/Analysis_of_variance).

3.1 ONE-WAY ANOVA

In statistics, *one-way analysis of variance* is a technique used to compare means of two or more samples (using the F-distribution). This technique can be used only for numerical data.

The ANOVA tests the null hypothesis that samples in two or more groups are drawn from populations with the same mean values. To do this, two estimates are made of the population variance. These estimates rely on various assumptions. The ANOVA produces an F-statistic, the ratio of the variance calculated between the means to the variance within the samples. If the group means are drawn from populations with the same mean values, the variance between the group means should be lower than the variance of the samples, following the central limit theorem. A higher ratio therefore implies that the samples were drawn from populations with different mean values.

Typically, however, the one-way ANOVA is used to test for differences between at least three groups, since the two-group case can be covered by a t-test. When there are only two means to compare, the t-test and the F-test are equivalent. A one-way ANOVA summary table is shown in Table 3.1.

The results of a one-way ANOVA can be considered reliable as long as the following assumptions are met:

- Response variable residuals are normally distributed.
- Samples are independent.

Table 3.1 One-way ANOVA summary table.

Source of variation	*SS*	df	*MS*	*F* calculated	*F* critical
Between groups					
Within groups (error)					
Total					

- Variances of populations are equal.
- Data are interval or nominal.

(http://en.wikipedia.org/wiki/One-way_ANOVA)

The hypothesis statements for a one-way ANOVA are as follows:

$$\mathrm{H}_0 : \mu_1 = \mu_2 = \mu_3 \ \ldots \ \mu_k \text{ (Null hypothesis)}$$

$$\mathrm{H}_\mathrm{A} : \text{The means are not equal (Alternative hypothesis)}$$

$$SS_{\mathrm{Total}} = \Sigma X_N^2 - \frac{(\Sigma X_N)^2}{N}$$

$$SS_{\mathrm{Between}} = \frac{(\Sigma X_1)^2}{n_1} + \frac{(\Sigma X_2)^2}{n_2} + \ldots + \frac{(\Sigma X_k)^2}{n_k} - \frac{(\Sigma X_N)^2}{N}$$

$$SS_{\mathrm{Within}} = SS_{\mathrm{Total}} - SS_{\mathrm{Between}}$$

$$\mathrm{df}_{\mathrm{Total}} = N - 1$$

$$\mathrm{df}_{\mathrm{Between}} = k - 1$$

$$\mathrm{df}_{\mathrm{Within}} = N - k$$

$$MS_{\mathrm{Between}} = \frac{SS_{\mathrm{Between}}}{\mathrm{df}_{\mathrm{Between}}}$$

$$MS_{\mathrm{Within}} = \frac{SS_{\mathrm{Within}}}{\mathrm{df}_{\mathrm{Within}}}$$

$$F = \frac{MS_{\mathrm{Between}}}{MS_{\mathrm{Within}}}$$

where

N = Total number of observations

K = Number of treatments

F_{Critical} = α, df, df values of the F-distribution (See Appendix B, Percentages of the F-Distribution)

Example: An engineer is unsure if a process is sensitive to a pressure setting. The engineer decides to conduct an ANOVA using three pressure settings: 100 psi, 110 psi, and

120 psi. The outputs are recorded below. We wish to test whether a change in pressure has an effect on the output with an α of 0.05.

Pressure (psi)		
100	110	120
8	5	2
9	4	4
8	5	5
7	7	2
9	5	7

$$SS_{\text{Total}} = \Sigma X_N^2 - \frac{(\Sigma X_N)^2}{N} = 577 - \frac{87^2}{15} = 72.4$$

$$SS_{\text{Between}} = \frac{(\Sigma X_1)^2}{n_1} + \frac{(\Sigma X_2)^2}{n_2} + \frac{(\Sigma X_3)^2}{n_3} = \frac{41^2}{5} + \frac{26^2}{5} + \frac{20^2}{5} = 46.8$$

$$SS_{\text{Within}} = SS_{\text{Total}} - SS_{\text{Between}} = 72.4 - 46.8 = 25.6$$

$$\text{df}_{\text{Total}} = N - 1 = 15 - 1 = 14$$

$$\text{df}_{\text{Between}} = k - 1 = 3 - 1 = 2$$

$$\text{df}_{\text{Within}} = N - k = 15 - 3 = 12$$

$$MS_{\text{Between}} = \frac{SS_{\text{Between}}}{\text{df}_{\text{Between}}} = \frac{46.8}{2} = 23.4$$

$$MS_{\text{Within}} = \frac{SS_{\text{Within}}}{\text{df}_{\text{Within}}} = \frac{25.6}{12} = 2.1$$

$$F = \frac{MS_{\text{Between}}}{MS_{\text{Within}}} = \frac{23.4}{2.1} = 10.97$$

The results are shown in Table 3.2. Since *F* calculated is greater than *F* critical at the 0.05 α (alpha) level, there is sufficient evidence to reject the null hypothesis $H_0 : \mu_1 = \mu_2 = \mu_3$ in favor of the alternative hypothesis that H_A : The means are not equal. See Figure 3.1.

Table 3.2 One-way ANOVA summary data table.

Source of variation	*SS*	df	*MS*	*F* calculated	*F* critical
Between groups	46.8	2	23.4	10.97	3.89
Within groups (error)	25.6	12	2.1		
Total	**72.4**	**14**			

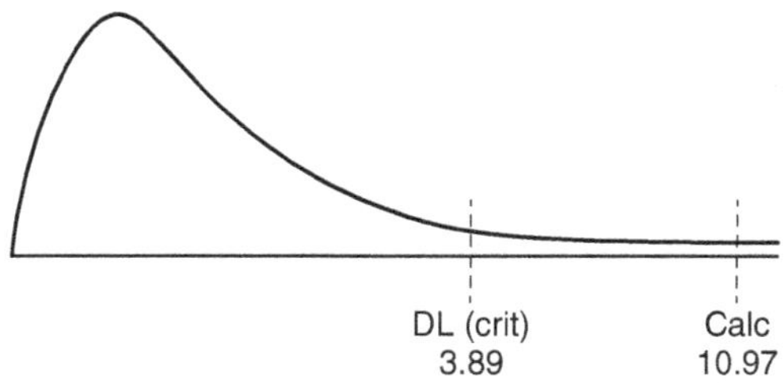

Figure 3.1 Decision limit.

Table 3.3 Two-way ANOVA summary table.

Source of variation	*SS*	df	*MS*	*F* calculated	*F* critical
Rows					
Columns					
Rows × Columns (interaction)					
Within (error)					
Total					

3.2 TWO-WAY ANOVA

In statistics, the *two-way analysis of variance* (ANOVA) test is an extension of the one-way ANOVA test. It examines the influence of different categorical independent variables on one dependent variable. While the one-way ANOVA measures the significant effect of one independent variable, the two-way ANOVA is used when there are more than one independent variable and multiple observations for each independent variable. The two-way ANOVA can not only determine the main effect of the contributions of each independent variable, but also identifies whether there is a significant interaction effect between the independent variables. A two-way ANOVA summary table is shown in Table 3.3.

As with other parametric tests, we make the following assumptions when using two-way ANOVA:

- Response variable residuals are normally distributed.
- Samples are independent.
- Variances of populations are equal.
- Data are interval or nominal.

(http://en.wikipedia.org/wiki/Two-way_analysis_of_variance)

The hypothesis statements for a two-way ANOVA are as follows:

$$H_0 \text{ (Rows)} : \mu_1 = \mu_2 = \mu_3 \ldots \mu_k \text{ (Null hypothesis)}$$

$$H_0 \text{ (Columns)} : \mu_1 = \mu_2 = \mu_3 \ldots \mu_k \text{ (Null hypothesis)}$$

$$H_0 \text{ (Interactions)} : \mu_1 = \mu_2 = \mu_3 \ldots \mu_k \text{ (Null hypothesis)}$$

$$H_A : \text{The means are not equal (Alternative hypothesis)}$$

$$SS_{\text{Total}} = \Sigma X_N^2 - \frac{(\Sigma X_N)^2}{N}$$

$$SS_{\text{Between}} = \frac{(\Sigma X_1)^2}{n_1} + \frac{(\Sigma X_2)^2}{n_2} + \ldots + \frac{(\Sigma X_k)^2}{n_k} - \frac{(\Sigma X_N)^2}{N}$$

Please note that this formula requires data from each individual cell. For example, C1-R1, C2-R1, C3-R1, C1-R2, C2-R2, C3-R2.

$$SS_{\text{Within}} = SS_{\text{Total}} - SS_{\text{Between}}$$

$$SS_{\text{Row}} = \Sigma \frac{(\Sigma_{\text{For each row}})^2}{n_{\text{For each row}}} - \frac{(\Sigma X_N)^2}{N}$$

$$SS_{\text{Column}} = \Sigma \frac{(\Sigma_{\text{For each column}})^2}{n_{\text{For each column}}} - \frac{(\Sigma X_N)^2}{N}$$

$$SS_{\text{Interaction}} = SS_{\text{Between}} - SS_{\text{Row}} - SS_{\text{Column}}$$

$$\text{df}_{\text{Total}} = N - 1$$

$$\text{df}_{\text{Row}} = r - 1$$

$$\text{df}_{\text{Column}} = c - 1$$

$$\text{df}_{\text{Row} \times \text{Column}} = (r - 1)(c - 1)$$

$$\text{df}_{\text{Within}} = N - rc$$

$$MS_{\text{Within}} = \frac{SS_{\text{Within}}}{\text{df}_{\text{Within}}}$$

$$MS_{\text{Row}} = \frac{SS_{\text{Row}}}{\text{df}_{\text{Row}}}$$

$$MS_{\text{Column}} = \frac{SS_{\text{Column}}}{\text{df}_{\text{Column}}}$$

$$MS_{\text{Row} \times \text{Column}} = \frac{SS_{\text{Row} \times \text{Column}}}{\text{df}_{\text{Row} \times \text{Column}}}$$

$$F_{\text{Row}} = \frac{MS_{\text{Row}}}{MS_{\text{Within}}}$$

$$F_{\text{Column}} = \frac{MS_{\text{Column}}}{MS_{\text{Within}}}$$

$$F_{\text{Row} \times \text{Column}} = \frac{MS_{\text{Row} \times \text{Column}}}{MS_{\text{Within}}}$$

where

N = Total number of observations

r = Number of rows

c = Number of columns

$F_{\text{Critical}} = \alpha$, df, df values of the F-distribution (See Appendix B, Percentages of the F-Distribution)

Example: An engineer is unsure whether a process is sensitive to a change in presses and dwell time. The engineer decides to conduct an ANOVA using three dwell times: 1.0 second, 1.5 seconds, and 2.5 seconds using press 1 and press 2. The outputs are recorded below. We wish to test whether a change in pressure has an effect on the output with an α of 0.05.

	Dwell time		
	1.0	**1.5**	**2.5**
Press 1	8	5	2
	9	4	4
	8	5	5
Press 2	7	7	2
	9	5	7
	7	6	5

$$SS_{\text{Total}} = \Sigma X_N^2 - \frac{(\Sigma X_N)^2}{N} = 687 - \frac{105^2}{18} = 74.5$$

$$SS_{\text{Between}} = \frac{(\Sigma X_1)^2}{n_1} + \frac{(\Sigma X_2)^2}{n_2} + \ldots + \frac{(\Sigma X_k)^2}{n_k} - \frac{(\Sigma X_N)^2}{N}$$

$$SS_{\text{Between}} = \frac{25^2}{3} + \frac{14^2}{3} + \frac{11^2}{3} + \frac{23^2}{3} + \frac{18^2}{3} + \frac{14^2}{3} - \frac{105^2}{18} = 51.2$$

$$SS_{\text{Within}} = SS_{\text{Total}} - SS_{\text{Between}} = 74.5 - 51.2 = 23.3$$

$$SS_{\text{Row}} = \Sigma \frac{(\Sigma_{\text{For each row}})^2}{n_{\text{For each row}}} - \frac{(\Sigma X_N)^2}{N} = \frac{50^2}{9} + \frac{55^2}{9} - \frac{105^2}{18} = 1.4$$

$$SS_{\text{Column}} = \Sigma \frac{(\Sigma_{\text{For each column}})^2}{n_{\text{For each column}}} - \frac{(\Sigma X_N)^2}{N} = \frac{48^2}{6} + \frac{32^2}{6} + \frac{25^2}{6} - \frac{105^2}{18} = 46.3$$

$$SS_{\text{Interaction}} = SS_{\text{Between}} - SS_{\text{Row}} - SS_{\text{Column}} = 51.2 - 1.4 - 46.3 = 3.5$$

$$\text{df}_{\text{Total}} = N - 1 = 18 - 1 = 17$$

$$\text{df}_{\text{Row}} = r - 1 = 2 - 1 = 1$$

$$\text{df}_{\text{Column}} = c - 1 = 3 - 1 = 2$$

$$\text{df}_{\text{Row} \times \text{Column}} = (r - 1)(c - 1) = (2 - 1)(3 - 1) = 2$$

$$\text{df}_{\text{Within}} = N - rc = 18 - 2 \times 3 = 12$$

$$MS_{\text{Within}} = \frac{SS_{\text{Within}}}{\text{df}_{\text{Within}}} = \frac{23.3}{12} = 1.9$$

$$MS_{\text{Row}} = \frac{SS_{\text{Row}}}{\text{df}_{\text{Row}}} = \frac{1.4}{1} = 1.4$$

$$MS_{\text{Column}} = \frac{SS_{\text{Column}}}{\text{df}_{\text{Column}}} = \frac{46.3}{2} = 23.2$$

$$MS_{\text{Row} \times \text{Column}} = \frac{SS_{\text{Row} \times \text{Column}}}{\text{df}_{\text{Row} \times \text{Column}}} = \frac{3.5}{2} = 1.8$$

$$F = \frac{MS_{\text{Row}}}{MS_{\text{Within}}} = \frac{1.4}{1.9} = 0.74$$

$$F = \frac{MS_{\text{Column}}}{MS_{\text{Within}}} = \frac{23.2}{1.9} = 12.21$$

$$F = \frac{MS_{\text{Row} \times \text{Column}}}{MS_{\text{Within}}} = \frac{1.8}{1.9} = 0.95$$

Results are shown in Table 3.4.

Table 3.4 Two-way ANOVA summary data table.

Source of variation	***SS***	**df**	***MS***	***F* calculated**	***F* critical**
Rows	1.4	1	1.4	0.74	4.75
Columns	46.3	2	23.2	12.21	3.89
Rows × Columns (interaction)	3.5	2	1.8	0.95	3.89
Within (error)	23.3	12	1.9		
Total	**74.5**	**17**			

Results for rows: Since F calculated is less than F critical at the 0.05 α (alpha) level, there is insufficient evidence to reject the null hypothesis that H_0 (Rows) : $\mu_1 = \mu_2 = \mu_3 \ldots \mu_k$. See Figure 3.2.

Results for columns: Since F calculated is greater than F critical at the 0.05 α (alpha) level, there is sufficient evidence to reject the null hypothesis H_0 (Columns) : $\mu_1 = \mu_2 = \mu_3 \ldots \mu_k$ in favor of the alternative hypothesis that H_A : The means are not equal. See Figure 3.3.

Results for interactions: Since F calculated is less than F critical at the 0.05 α (alpha) level, there is insufficient evidence to reject the null hypothesis that H_0 (Interactions) : $\mu_1 = \mu_2 = \mu_3 \ldots \mu_k$. See Figure 3.4.

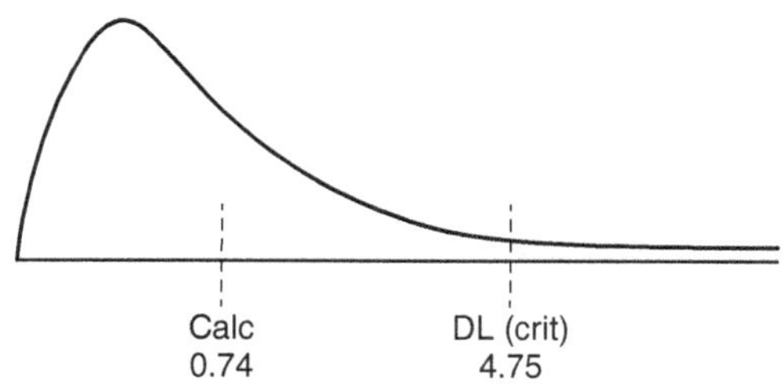

Figure 3.2 Decision limit for rows (pressure).

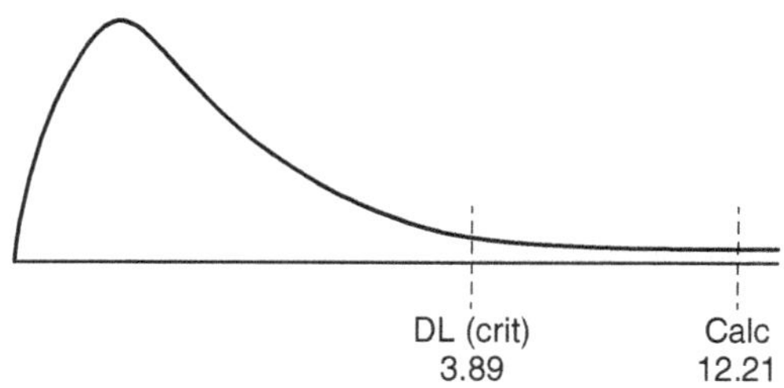

Figure 3.3 Decision limit for columns (dwell time).

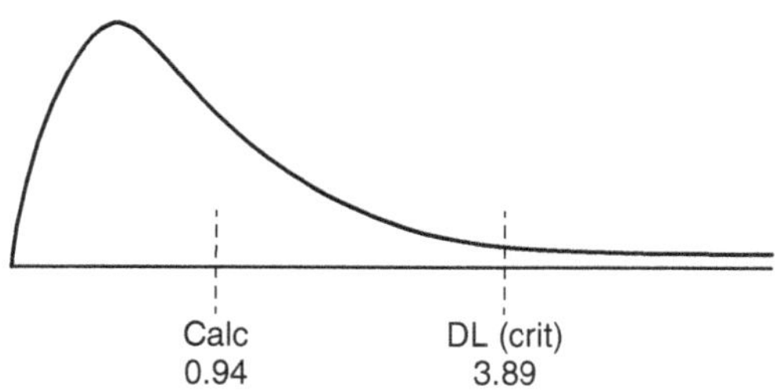

Figure 3.4 Decision limit interaction (pressure and dwell time).

4
Experiments with Two Factors

An experiment with two factors is presented to begin the process of defining terms and procedures. The techniques developed in this small experiment can then be used in experiments with many more factors. Each example will introduce additional concepts, building upon the techniques previously presented.

The first example will demonstrate the general concepts of planning and conducting a basic two-factor DOE and will introduce the concept of *interpolation*, which assumes a linear model. The second example builds upon the first example and will present a simple method to test the assumption of linearity.

4.1 BOND STRENGTH EXAMPLE

A manufacturer of laminated papers takes two rolls of kraft paper and extrudes a layer of polymer in between, simultaneously pressing the three components into a sandwich. The customer has complained about the lack of adherence of the polymer to the two paper layers. Therefore, the objective of the study is to maximize the bond strength. Note that the objective must be defined up front. The manufacturer has developed a test for this bonding using a specialized measuring device. This is the key output measure; in DOE, it is called the *response variable*. The team working on improving the bonding process has decided that the best candidate factors for controlling the bond strength are *polymer temperature* and *paper source*. *Factors* are process variables that can be controlled during the experiment. Think of them as machine settings that can be set at various levels. Such factors are also referred to as *independent variables* in some references. These factors were selected based on the review of data from the process and on a brainstorming exercise conducted by the team. After much discussion, the *levels* (the different options or settings for each factor in the experiment) were set as shown below.

Factor	**Low level**	**High level**
A. Polymer temperature °F	580	600
B. Paper source	Vendor Y	Vendor X

Note that factor A is a quantitative factor since its levels can be set along a relatively continuous measurement scale. Factor B is qualitative since its levels are discrete; that is, there are a finite number of levels available. Where possible, quantitative factors are preferred since that permits the experimenter to follow such a factor to an optimum condition with respect to the response.

Next, the experimental design must be defined. The *experimental design* is the definition of the collection of trials to be run in the experiment. In this example, the design is all possible combinations of the chosen factor levels, called a *full-factorial* design. Since each factor has two levels, this is a 2^2 design requiring four unique treatments, or runs. (The base refers to the number of levels; the exponent refers to the number of factors.) These four runs are the four combinations of the levels of the factors, using either the labels high and low or plus and minus signs to identify the actual levels:

1. A low; B low = A–B–
2. A low; B high = A–B+
3. A high; B low = A+B–
4. A high; B high = A+B+

A *run*, or treatment, is the unique combination of the factor levels. Note that each run may be performed more than once. The procedure of performing more than one trial of each run is referred to as *replication* when each experimental trial utilizes a completely new setup. A *replicate* is an independent and random application of the run, including the setup. This is considerably different from a *repeat*, which is a repetition of a run without going through a new setup. The 2^2 design defines four treatments, or runs. To reduce the impact of the inherent variation in the process, each run is replicated, for a total of eight trials. These eight trials must be carried out in a random order to minimize the risk of bias in the results due to unknown or uncontrolled factors.

Randomization refers to the order in which the trials of an experiment are performed. Randomization can be achieved by numbering the trials and then drawing numbers from a hat, using a table of random numbers, shuffling numbered cards, and so on. This is important to protect against uncontrolled and/or unknown influences of variables that are not part of the experiment. As an example, assume that an experiment has a single factor: pressure. Assume also that, unknown to the experimenters, a thermostat reading is drifting steadily downward and that the background temperature affects the response. If all the low-pressure trials are performed on day one and the high-pressure trials on day two, is their difference due to the change in pressure or to the change in temperature? The experimenters cannot be sure and, in fact, may not even know there is a problem. If the temperature impact is major, one could erroneously conclude that pressure is a causative factor. In order to minimize this risk of unknown influence, the experimenters randomly assign the order of testing to improve the chances of averaging out this bias or distortion of the responses related to the factor(s) under study.

There are times when one cannot randomize due to physical or cost constraints. Such cases generally will require blocking of the experiment into sections defined by the factor that cannot be randomized. For instance, there may not be enough material from one batch of raw material to complete the experiment. Instead, one may block by carrying out a carefully defined set of trials with one batch and the remaining trials with

Table 4.1 Bond strength example data.

Run	A Temp	B Vendor	AB Interaction	Y^1	Y^2	$\bar{Y}$	s^2	s
1	–	–	+	18.60	17.40	18.00	0.72	0.85
2	+	–	–	17.50	16.50	17.00	0.50	0.71
3	–	+	–	18.20	16.70	17.45	1.13	1.06
4	+	+	+	22.90	22.20	22.55	0.25	0.50
$\Sigma\bar{Y}+$	39.55	40.00	40.55					
$\Sigma\bar{Y}-$	35.45	35.00	34.45			$\bar{\bar{Y}}$ 18.75		
$\bar{Y}+$	19.78	20.00	20.28					
$\bar{Y}-$	17.73	17.50	17.23					
C+	1.03	1.25	1.53					
C–	–1.02	–1.25	–1.52					
Abs effect	2.05	2.50	3.05					

a second batch. Such blocking minimizes the risk of the nuisance-factor batches creating excessive estimates of the inherent variation.

This first part of the chapter is the design part of DOE: factor selection, setting levels, defining treatments, randomization of the order of performance. Now the eight trials must be performed exactly as required, keeping all other factors constant. The responses or results that were obtained are shown in Table 4.1.

Before we begin, an explanation of the table is in order. The format of this table will be utilized throughout this text. However, there will be times when it will vary slightly, as in the case of attribute and signal-to-noise (S/N) outputs. The use of plus and minus signs is way to indicate the high and low levels for each factor. The various fields will be further explained during the progression of the following example.

Analysis of these data will be done using the nine-step analytical procedure that follows. These nine steps will work for experiments with any number of factors, as long as the factors have two levels. At first glance, the limitation to two levels may seem very restrictive. In actuality, it permits a remarkable efficiency in the number of trials needed. The primary penalty is that the relationship between a response and a factor with two levels is assumed to be linear. Techniques for identifying nonlinear relationships (curvature) are presented later.

Nine Steps for Analysis of Effects

1. *Calculate the absolute values of the effects. Main effects* are defined as the difference in the average response between the high (+1) and low (–1) levels of a factor (see Table 4.1 for the data used in this example). The effect of A is written as $E(A)$.

Using plus and minus signs to represent high and low levels of a factor, main effects are defined as

$$E(A)=|\bar{Y}_{A+}|-|\bar{Y}_{A-}|=\frac{22.55+17.00}{2}-\frac{17.45+18.00}{2}=2.05.$$

Note: The minus signs have been removed for simplicity.

The main effect of +2.05 means that the average bond strength at the high level of temperature (600 °F) was 2.05 units higher than the average bond strength at the low temperature (580 °F). The same data can now be used to determine the main effect of factor B (vendor) on the process. Note that a plus sign is used to designate vendor X while a minus sign is used to designate vendor Y. This is an arbitrary choice that once chosen must be used consistently.

$$E(B)=|\bar{Y}_{B+}|-|\bar{Y}_{B-}|=\frac{22.55+17.45}{2}-\frac{17.00+18.00}{2}=2.50.$$

Note: The minus signs have been removed for simplicity.

This indicates that the paper supplied by vendor X averaged 2.50 units higher in bond strength than that supplied by vendor Y. An *effect* is the difference in averages. Note that in a DOE the same data are used for more than one factor. This is the fundamental concept of DOE: an experiment is used to define many effects, rather than performing an experiment for each factor. The graphs in Figures 4.1 and 4.2 indicate the direction of influence for the factors and communicate the meaning of the effects very clearly.

The plots of the main effects of temperature and vendor indicate that a higher temperature results in a stronger bond, and that vendor X is superior to vendor Y. Observe

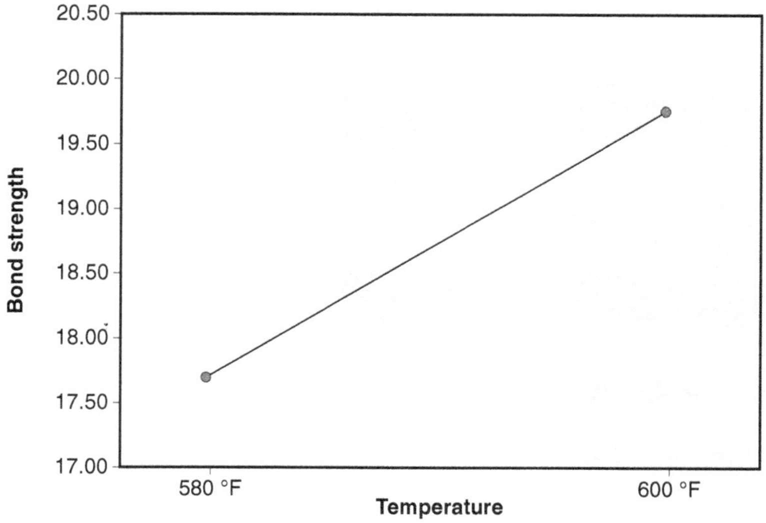

Figure 4.1 Plot of effect A (temperature) on bond strength.

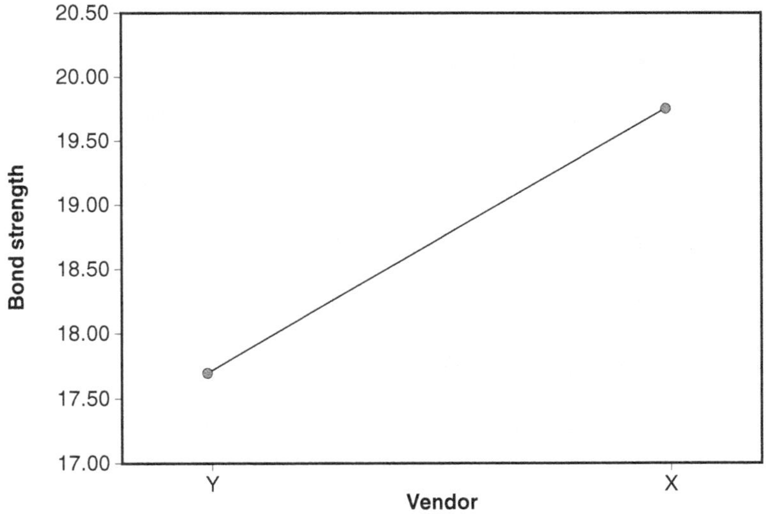

Figure 4.2 Plot of effect B (vendor) on bond strength.

the data in Table 4.1. Are those results always true and to the amount expected? Are there certain combinations of temperature and vendor that don't seem to provide the expected results? An interaction occurs when a particular combination of two factors indicates something unexpected from simply observing their main effects.

$$E(AB)=\left|\bar{Y}_{AB+}\right|-\left|\bar{Y}_{AB-}\right|=\left(\frac{18.00+22.55}{2}-\frac{17.00+17.45}{2}\right)=3.05$$

Note: The minus signs have been removed for simplicity.

The interaction effect essentially increases or decreases the main effect in the experiment by 3.05 units of strength. For example, the main effect of temperature equals +2.05. However, when paper from vendor X is used, the effect of temperature is actually 5.10, that is, (22.55 – 17.45). Conversely, when paper from vendor Y is used, the temperature effect is –1.0, that is, (17.0 – 18.0). This means that when using paper from vendor Y, higher bond strengths are achieved with lower temperature. The plot reveals what the interaction effect means (see Figures 4.3 and 4.4).

Interactions can be plotted as temperature-vendor (as shown in Figure 4.3) or as vendor-temperature; both are correct and are simply two views of the same phenomenon.

> A final precaution on interactions: if an interaction proves to be significant, the interaction chart is more important than the main effect charts. If there is an interaction, the main effect describes average results whereas the interaction is more appropriate in describing the joint effect.

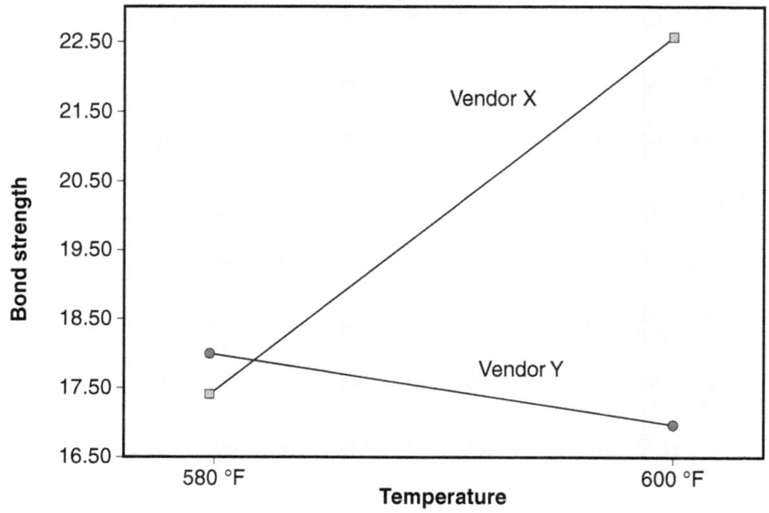

Figure 4.3 Plot of interaction AB (temperature-vendor).

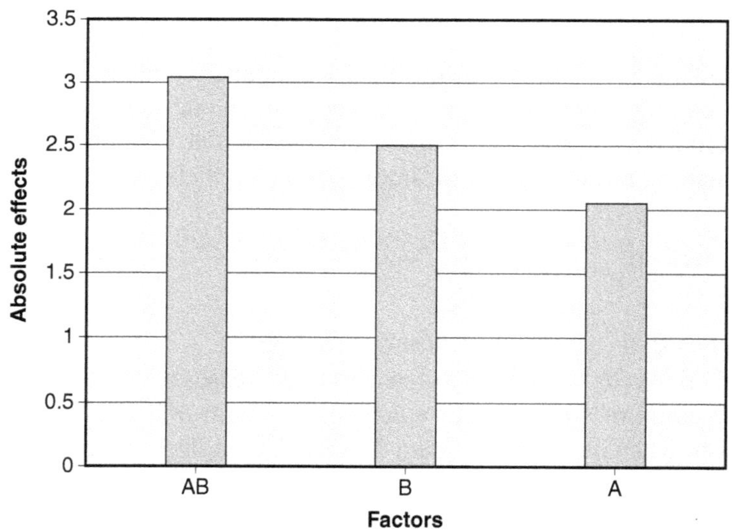

Figure 4.4 Pareto chart of the absolute values of the effects.

2. *Create a Pareto chart of the absolute values of the effects.* To show the relative importance of the effects, plot their absolute values with a Pareto chart (see Figure 4.4). The Pareto chart, which becomes more useful when larger experiments are studied, is often sufficient to determine what effects are meaningful. The problem now is to test these effects for significance; that is, how do we show that the results of the experiment are beyond the inherent variation to be expected in the experiment? The next steps will determine whether these three effects are significant or meaningful.

3. *Calculate the standard deviation of the experiment* (s_e). Determination of significance requires calculation of a standard deviation as a measure of the inherent variation, or experimental error, in the process. To obtain s, the variance (s^2) is calculated for each run, or treatment. These variances are then averaged and converted to a standard deviation (s) by taking the square root. Variances are averaged even though standard deviations are the statistic used. This is to permit creation of a measure of spread or variation that will always be a positive number. The variances are calculated for each run and averaged. The average variance is converted to a standard deviation by taking the square root using the following formula:

$$s_e = \sqrt{\left(\Sigma s^2\right)/k}$$

where

k = The number of runs

s^2 = The variance from each run.

For this example,

$$s_e = \sqrt{\left(\Sigma s^2\right)/k} = \sqrt{(0.72+0.50+1.13+0.25)/4} = 0.81.$$

This estimate of the inherent variation represents (1) measurement variation, plus (2) the inability of the experimenter to repeat the conditions, plus (3) the inability of the process to repeat the same response for the same conditions.

4. *Calculate the standard deviation of the effects* (s_{Eff}). *Effects* are differences between averages, requiring a modified standard deviation. This is called the *standard deviation of the effects* and is defined as

$$s_{\text{Eff}} = s_e\sqrt{4/n}$$

where

n = The total number of trials (number of runs × number of replicates)

s_e = The standard deviation.

This formula will hold for any number of trials as long as the factors have two levels. For this example,

$$s_{\text{Eff}} = s_e\sqrt{4/n} = 0.81\sqrt{4/8} = 0.57$$

5. *Determine the* t*-statistic.* To use the t-table, the degrees of freedom in the experiment must be determined. Degrees of freedom (df) measure the amount of information available to estimate the standard deviation. The calculation is

df = (# of observations per run – 1) × (# of runs).

Note: If the # of observations per run – 1 = 0, use a multiplier of 1.

For this example,

$$df = (\text{\# of observations per run} - 1) \times (\text{\# of runs}) = (2 - 1) \times (4) = 4$$

$$t_{\alpha/2}, df.$$

Next, we must calculate a critical value to determine the significance of the effects. Confidence is 1– alpha risk, where the alpha risk is the chance of erroneously claiming significance. The typical alpha risk (also written with the Greek symbol α) is therefore 5 percent. Referring to Appendix C, Percentage Points of the Student's *t*-Distribution, for 4 degrees of freedom and α/2, the critical value is $t = 2.776$:

$$t_{\alpha/2,df} = t0.050/2{,}4 = 2.776.$$

6. *Calculate the decision limits.* The *decision limits* (DL) for the significance of effects in this DOE can now be calculated. The question is whether the computed effects are significantly different from zero and, therefore, not due to random variation. If the effects are outside the zone defined by the decision limits, the effects are considered real or significant. The decision limits are calculated by

$$DL = (t_{\alpha/2,df})\,(s_{Eff}) = (2.776)\,(0.57) = 1.58.$$

The results are shown graphically in Figure 4.5.

7. *Determine the significant effects.* Both of the effects (A and B) and the interaction (AB) exceed the decision limit so are judged to be real or significant. Since the interaction effect was significant, one must jointly consider the two factors to determine how to optimize the process.

A half-normal probability plot may not be appropriate for this example because the number of effects is small (two-factor experiment with one interaction). Due to the limited amount of information available, the results using a half-normal probability plot may yield misleading information.

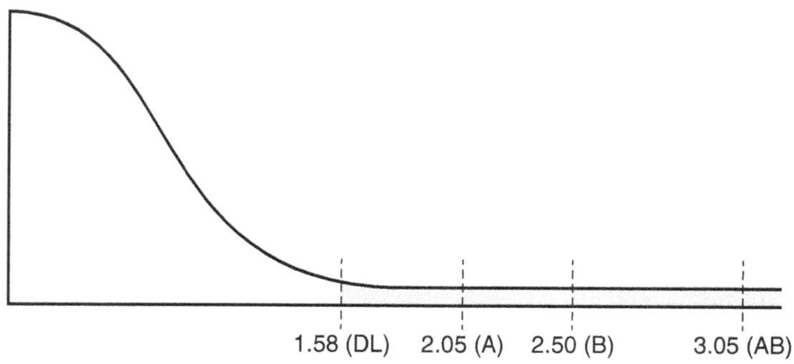

Figure 4.5 Decision limits for the effects and interactions.

8. *Graph the main effects and any significant interactions.* In reality, graphs need only be made for significant effects. If an effect is not significant, the graph is of no interest. If an interaction is significant, that graph is more meaningful than those of the individual factors since an interaction means that the experimenter must consider the factors jointly. The graphs for this example have already been shown in Figures 4.1, 4.2, 4.3, and 4.4.

9. *Model the significant effects, significant interactions, and individual effects from significant interactions.* The model or prediction equation is useful to predict the optimum outcome for future validation experiments. The model is a linear equation of the following form, using only the significant effects. *However, if an interaction effect is significant, the terms for the two main effects are also included even if they are not significant.* This is due to the hierarchy rule for defining a model. This rule becomes more important in advanced techniques for optimization.

> If an interaction effect is significant, the terms for the two main effects are also included even if they are not significant.

The coefficients must be calculated so the model can be compiled. The coefficient calculations are simply the average difference of the overall average from a level (+, –). Note that it is not necessary to calculate all of the coefficients. All coefficients are calculated for illustrative purposes.

$$C_{+,-} = \bar{Y}_{+,-} - \bar{\bar{Y}}$$

where

$\bar{\bar{Y}}$ = Average run S/N ratio

$\bar{Y}$ = Average at a treatment level

$A_+ = 19.78 - 18.75 = 1.03$

$A_- = 17.73 - 18.75 = -1.02$

$B_+ = 20.00 - 18.75 = 1.25$

$B_- = 17.50 - 18.75 = -1.25$

$AB_+ = 20.28 - 18.75 = 1.53$

$AB_- = 17.23 - 18.75 = -1.52$

Since our objective is to maximize bond strength, A (polymer temperature 600 °F) and B (paper source vendor X) are set at the high (+1) levels that will provide maximum strength. Refer to Figure 4.3.

$$\hat{Y} = \bar{\bar{Y}} + \text{coefficients of the significant effects, significant interactions, and individual effects from significant interactions}$$

$$\hat{Y} = \bar{\bar{Y}} + A_+ + B_+ + AB_+$$

$$\hat{Y} = 18.75 + 1.03 + 1.25 + 1.53 = 22.56$$

If the best combination of temperature and vendor paper is used, bond strength is estimated to average 22.56. This is better than the results predicted simply from the main effects of A and B due to the contribution of the interaction term.

> An interaction either increases the response over that expected due to main effects alone or it reduces the response below that expected from main effects.

At this point it would be wise to verify experimental results by using the model of the experiment to operate the process. All nonsignificant factors should be set solely on the basis of cost, productivity, or convenience. One may also consider opening up the window of the significant factors to see if further process improvement is possible.

What if one needed to know the expected response for a temperature of 595 °F with paper from vendor X? The model uses coded values for the levels, so the temperature of 595 °F must be changed to coded values to interpret the results (see Figure 4.6).

Since 1 unit = 10 °F and 595 °F is 5 degrees from the zero point equivalent to 590 °F, the coded equivalent of 595 °F is 1/2 = + 0.5. This value is then inserted into the model:

$$\hat{Y} = \bar{\bar{Y}} + (0.5)A_+ + B_+ + (0.5)AB_+$$

$$\hat{Y} = 18.75 + (0.5)1.03 + 1.25 + (0.5)1.53 = 21.28$$

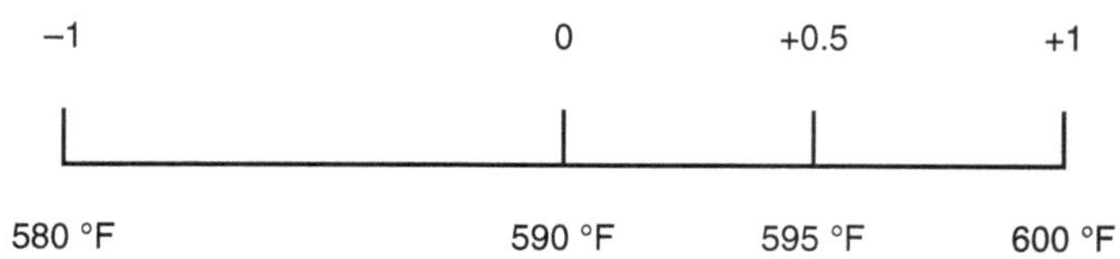

Figure 4.6 Temperature scale range.

This interpolation procedure is useful when there are only a couple of significant factors. Otherwise, too many assumptions must be made since one would have a single equation in several unknowns.

What further study can be recommended if even higher bond strength is desired? Does the temperature effect continue beyond 600 °F? It is dangerous to extrapolate beyond the range of study (580 °F to 600 °F). Plan a verification study at 600 °F and at 610 °F with paper from vendor X. This will verify the previous conclusions and explore the potential for even better performance.

The objective in this example was to achieve stronger bond strength. Suppose the objective had been different. What if the bond strength problem was not one of inadequate strength but of too much variation in the strength since temperature could not be tightly controlled? What conclusions would be drawn in that circumstance? What conditions would make the process robust or insensitive to changes in the process variables of temperature and vendor? The interaction plots show that changes in temperature have little effect on bond strength if paper from vendor Y is used. This is why the objective of the study must be clearly understood before the experiment is finalized.

For demonstration purposes, again assume a different objective. Suppose that, instead of achieving a maximum or minimum, the experimenter needed to hit a target response of 20. This would lead one to determine the setting of the quantitative (numeric) factor of temperature. Based on the interaction graph (Figure 4.3), note that only vendor X material provides a range of responses that includes the value of 20. If B is set at +1 for vendor X, what setting for factor A (temperature) would provide a forecast of the target value? To determine this, solve the model for A where B = +1 (vendor X) and Y = 20 (target response):

$$\hat{Y} = \bar{\bar{Y}} + A_+ + B_+ + AB_+$$

$$20 = 18.75 + 1.03 + 1.25 + 1.53$$

$$0 = 2.56A$$

A = 0 in coded units or, by interpolation, A = 590 °F in actual units.

This completes the analytical procedure. It is very important that the conclusions from this experiment be verified either by operating at the recommended conditions or by further experimentation. All reports, graphs, and recommendations should be put in layman's terms, not in the statistical terminology that is used in the analysis. Remember, your audience may not know what A means, but may understand Pareto charts (the 80/20 rule) and graphs as long as the references are in familiar language.

The steps provided above are the fundamental analysis techniques for any number of factors at two levels. Remember that small DOEs such as the previous example are usually used in refining experiments after larger screening experiments or when some quick troubleshooting is needed. The spacing of the quantitative levels (*temperature* in our example) is also important. In a screening design, the levels should be set as wide or as bold as practical to make it easier to discover significant factors for later study. In a refining design, the levels would be set more closely and more replication would be required since the size of the effects would be reduced. Additional data would be needed to demonstrate significance of the smaller effects.

A final caution is needed on the statistical control of the process. Lack of control increases the experimental error and can also create false effects. It is important to stabilize the process, that is, eliminate the special causes. Otherwise, replication must be increased to overcome the distortion (overestimate) of the estimate of inherent variation (s_e) due to the presence of special causes of variation.

4.2 NONLINEAR MODELS

Restricting the experiments to two levels allows simplification of the analyses and provides substantial reduction in the number of runs required in larger experiments. However, this permits only linear (straight-line) relationships to be defined. This is generally sufficient for most experimental work until the final refining or optimization stages of a study. At that time, three or more levels may be used to identify a nonlinear or quadratic relationship. Center points can also be used to detect the presence of nonlinearity or curvature. These are trials run at the midpoint of the factor levels—the zero level. (Obviously, nonlinearity only has meaning with quantitative factors.) Center points also provide an opportunity to increase the degrees of freedom and improve the estimate of the experimental error (s_e). The following procedures apply in addressing nonlinearity:

1. Modify the procedure for estimating the experimental error.

   a. Calculate and average the variances of the base (non–center point) runs as usual. Then calculate the variance of the center points. Define these variances as s_b^2 and s_c^2, respectively.

   b. Calculate the degrees of freedom for the base design and for the center points.

      Degrees of freedom for the base design = dfb =
      (Number of runs) × (Number of replicates – 1)

      Degrees of freedom for the center points = dfc =
      (Number of center points – 1)

   c. Calculate s_e^2 as the weighted average of the two variance estimates, weighting each by its degrees of freedom:

$$s_e^2 = \frac{\left(\left(\mathrm{df}_b \times s_b^2\right) + \left(\mathrm{df}_c \times s_c^2\right)\right)}{\left(\mathrm{df}_b + \mathrm{df}_c\right)}$$

$$s_e = \sqrt{s_e^2}$$

2. Define the nonlinear effect.

$$E(\text{nonlinearaity}) = \bar{\bar{Y}} - \bar{Y}_{\text{Center}}$$

where

$\bar{\bar{Y}}$ = The average of all trials except center points

$\bar{Y}_{\text{Center}}$ = The average of the center points.

If the relationships are linear, the center points should produce an average response that approximates the grand average ($\bar{\bar{Y}}$). If the average of the responses at the center point is significantly different from the grand average, nonlinearity exists.

3. Define the standard deviation to test nonlinearity:

$$s(\text{nonlinear}) = s_e\sqrt{\frac{1}{n}+\frac{1}{c}}$$

where

n = The total number of trials not at the center

c = The number of trials at the center.

4. Determine the t-statistic using all the degrees of freedom:

$$\text{Degrees of freedom} = \text{df}_b + \text{df}_c.$$

This was the reason for adding the additional calculations using center points. The extra degrees of freedom will provide a stronger t-statistic. This t-statistic can be used for all effects, nonlinear and basic.

5. Calculate a decision limit for the nonlinear effect.

$$\text{DL} = \pm\,(t_{\alpha/2,\text{df}}) \times s_{\text{Nonlinear}}$$

The following guidelines must be remembered when dealing with nonlinearity:

1. The presence of center points changes the calculation of s_e, degrees of freedom, and the t-statistic for all assessments. The increased complexity is the price of increasing the degrees of freedom. Since the expense of an experiment usually leads to minimum degrees of freedom, the additional complexity is worth it.

2. If the nonlinearity effect is significant, the linear plots for main effects are suspect. The test for nonlinearity does not tell the experimenter which factor(s) contain the curvature, only that it exists. Significance of main effects is no longer reliable.

3. Interpolation of the linear model is no longer wise.
4. Significant nonlinearity must be evaluated by further experimentation: (a) expanding the current experiment to three levels or (b) repeating the DOE from the beginning with three levels.
5. There should be at least four center points to increase the chance of detecting nonlinearity.

4.3 CORROSION STUDY EXAMPLE

A research group has been charged with developing steel with better corrosion resistance than the current product. Previous experimentation has established the chemical composition of all but two elements. Chromium and nickel levels have been narrowed but not finalized. The team wishes to evaluate (1) chrome with levels of 0.04 percent and 0.10 percent, and (2) nickel with levels of 0.10 percent and 0.20 percent. The design was a 22 factorial with three replicates. Since there is a concern over nonlinearity, five center points were run at the factor settings of Cr at 0.07 percent and Ni at 0.15 percent. The order of the seventeen experimental trials was randomized (the four treatments were each repeated three times, and the center point treatment was repeated five times), and the response is the weight loss in an acid bath during a fixed time period. The objective is to minimize the weight loss. Data are given in Table 4.2

Table 4.2 Corrosion study example data.

Treatment	A Chrome	B Nickel	AB Interaction	Y^1	Y^2	Y^3	$\bar{Y}$	s^2	s
1	–	–	+	4.10	4.90	4.30	4.43	0.173	0.416
2	+	–	–	5.50	4.00	4.30	4.60	0.630	0.794
3	–	+	–	9.60	8.70	10.10	9.47	0.503	0.709
4	+	+	+	8.20	7.40	8.70	8.10	0.430	0.656
$\Sigma\bar{Y}+$	12.7	17.57	12.53						
$\Sigma\bar{Y}-$	13.9	9.03	14.07				$\bar{\bar{Y}}$ 6.65		
$\bar{Y}+$	6.35	8.79	6.27						
$\bar{Y}-$	6.95	4.52	7.04						
C+	–0.30	2.14	–0.38						
C–	0.30	–2.13	0.39						
Abs effect	0.60	4.27	0.77						

Responses at center point: 9.5, 9.0, 8.6, 8.7, 8.1 ($\bar{Y}$ = 8.78, s^2 = 0.267).

Nine Steps for Analysis of Effects

1. *Calculate the absolute values of the effects*:

$$E(A)=\left|\overline{Y}_{A+}\right|-\left|\overline{Y}_{A-}\right|=\frac{4.60+8.10}{2}-\frac{4.43+9.47}{2}=0.60$$

$$E(B)=\left|\overline{Y}_{B+}\right|-\left|\overline{Y}_{B-}\right|=\frac{9.47+8.10}{2}-\frac{4.43+4.60}{2}=4.27$$

$$E(AB)=\left|\overline{Y}_{AB+}\right|-\left|\overline{Y}_{AB-}\right|=\left(\frac{4.43+8.10}{2}-\frac{4.60+9.47}{2}\right)=0.77.$$

Note: The minus signs have been removed for simplicity.

2. *Create a Pareto chart of the absolute values of the effects.* Figure 4.7 indicates that factor B (nickel) is significant. The following steps will determine whether any of the factors or interaction is significant.

3. *Calculate the standard deviation of the experiment* (s_e). Since there are center points, additional degrees of freedom and a better estimate of experimental error are possible by modifying the usual procedures. If there were no center points, the usual procedures would be applied.

   a. Calculate and average the variances of the base four runs. Then calculate the variance at the center point.

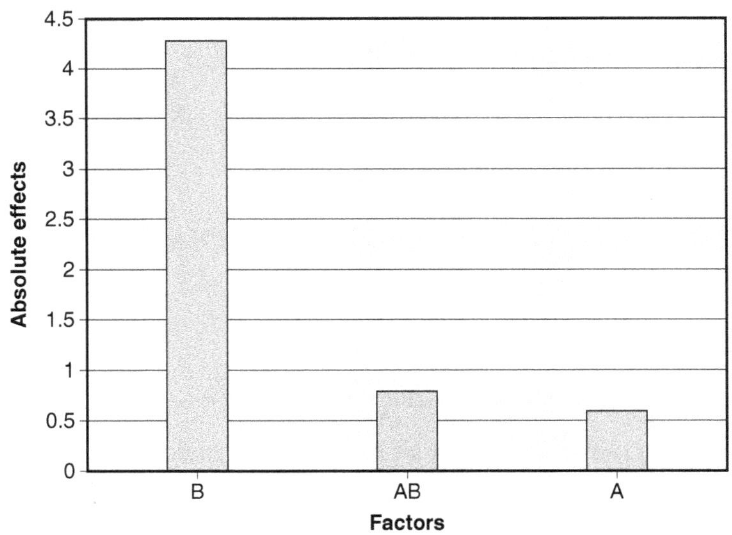

Figure 4.7 Pareto chart of the absolute values of the effects.

For the base,

$$s_b^2 = \frac{0.173+0.630+0.503+0.430}{4} = 0.434$$

For the center points,

$$s_c^2 = 0.267$$

b. Calculate the degrees of freedom for the base design and for the center points.

$$df_b = \text{(Number of runs)} \times \text{(Number of replicates} - 1) = 4 \times (3-1) = 8$$

$$df_c = \text{(Number of center points} - 1) = 5 - 1 = 4$$

Note: If the number of replicates – 1 = 0, use 1 as a multiplier.

c. Calculate s_e^2 and s_e as the weighted average of the two variance estimates, weighting each by its degrees of freedom.

$$s_e^2 = \frac{\left(\left(df_b \times s_b^2\right)+\left(df_c \times s_c^2\right)\right)}{\left(df_b + df_c\right)} = \frac{((8\times 0.434)+(4\times 0.267))}{(8+4)} = 0.378$$

$$s_e = \sqrt{s_e^2} = \sqrt{0.378} = 0.615$$

4. *Calculate the standard deviation of the effects* (s_{Eff}):

$$s_{\text{Eff}} = s_e\sqrt{4/n} = .615\sqrt{4/12} = .355$$

where

n = Total number of trials (Number of runs × Number of replicates)

s_e = Standard deviation.

5. *Determine the* t-*statistic*:

Total degrees of freedom = $(df_b + df_c) = 8 + 4 = 12$.

With a typical alpha risk (also written with the Greek symbol α) of 5 percent, referring to Appendix C, Percentage Points of the Student's t-Distribution for 12 degrees of freedom and $\alpha/2$, the critical value is $t = 2.179$.

6. *Calculate the decision limits* (Figure 4.8):

$$DL = (t_{\alpha/2,\text{df}})\,(s_{\text{Eff}}) = (2.179)\,(0.355) = 0.77$$

7. *Determine the significant effects.* The results shown in Figure 4.8 indicate that factor B (nickel) is significant.

A half-normal probability plot may not be appropriate for this example because the number of effects is small (two-factor experiment with one interaction). The results using this method may be erratic.

8. *Graph the main effects and any significant interactions* (Figure 4.9, Figure 4.10, and Figure 4.11). An analysis of Figures 4.9, 4.10, and 4.11 indicate that nickel should be set at the low (0.10 percent) level to minimize weight loss. Note that interaction AB is not significant and is shown for illustrative purposes only.

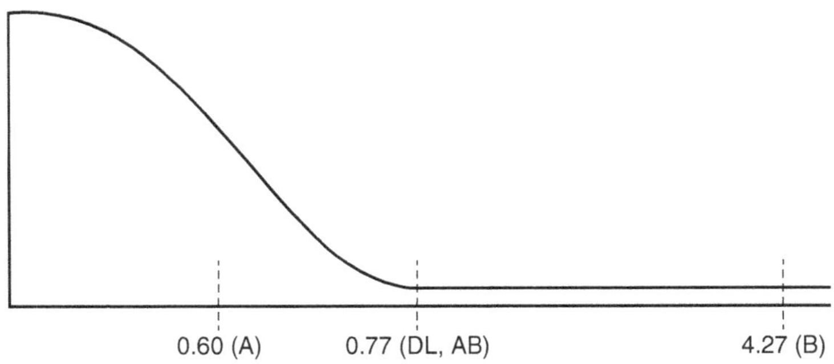

Figure 4.8 Decision limits for the effects.

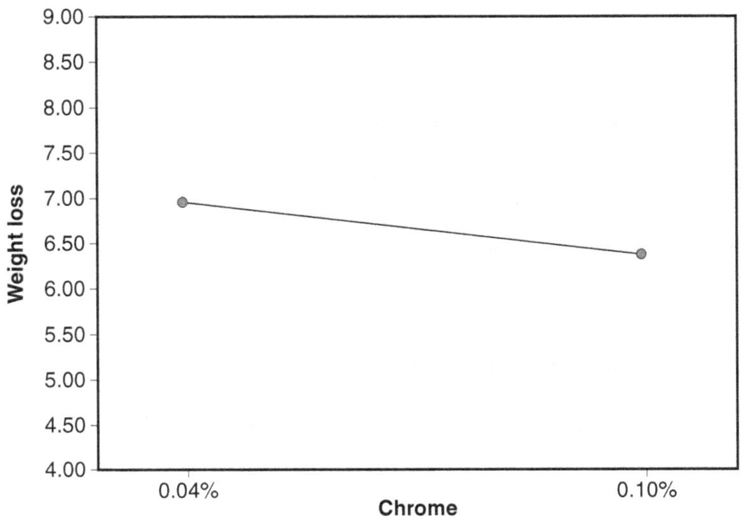

Figure 4.9 Plot of effect A (chrome) on on weight loss.

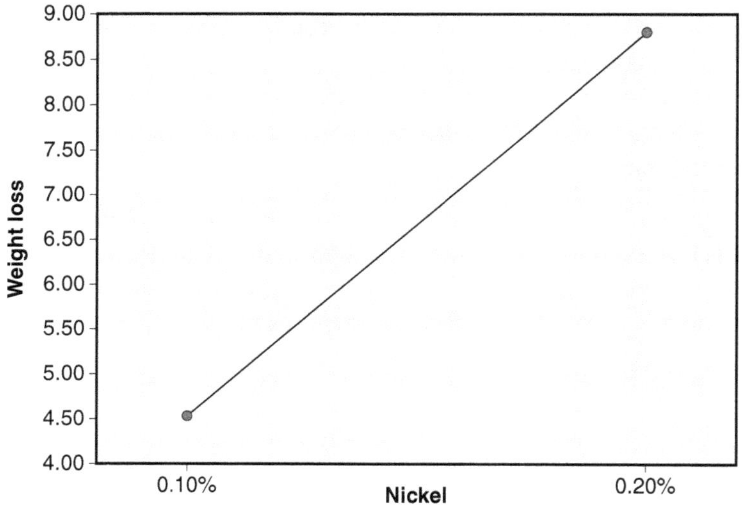

Figure 4.10 Plot of effect B (nickel) on on weight loss.

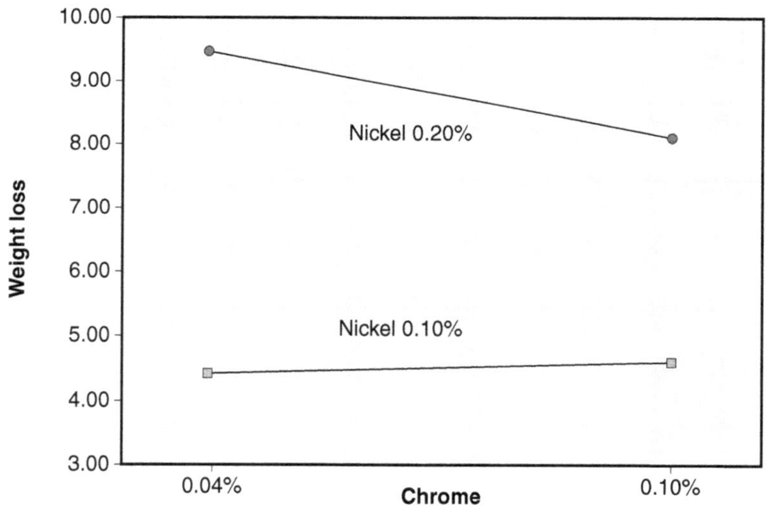

Figure 4.11 Plot of interaction AB (chrome-nickel) on weight loss.

9. *Model the significant effects, significant interactions, and individual effects from significant interactions.* The coefficients must be calculated so the model can be compiled. The coefficient calculations are simply the average difference of the overall average from a level (+, –). Note that it is not necessary to calculate all of the coefficients. All coefficients are calculated for illustrative purposes.

$$C_{+,-} = \overline{Y}_{+,-} - \overline{\overline{Y}}$$

where

$\overline{\overline{Y}}$ = Average run S/N ratio

$\overline{Y}$ = Average at a treatment level.

$A_+ = 6.35 - 6.65 = -0.03$

$A_- = 6.95 - 6.65 = 0.03$

$B_+ = 8.79 - 6.65 = 2.14$

$B_- = 4.52 - 6.65 = -2.13$

$AB_+ = 6.27 - 6.65 = -0.38$

$AB_- = 7.04 - 6.65 = 0.39$

Since our objective is to minimize weight loss, factor B (nickel) should be set at 0.01% (–1 level). Refer to Figure 4.10. Factor A (chrome) is not significant and can be set solely on the basis of cost, productivity, or convenience.

$\hat{Y} = \overline{\overline{Y}} +$ coefficients of the significant effects, significant interactions, and individual effects from significant interactions

$$\hat{Y} = \overline{\overline{Y}} + B_-$$

$$\hat{Y} = 6.65 + -2.13 = 4.52$$

If 0.10% nickel is used, the weight loss is estimated to average 4.52.
However, before continuing, the nonlinearity check must be made.

a. Define the effect of nonlinearity:

$$E(\text{Nonlinearity}) = \overline{\overline{Y}} - \overline{Y}_{\text{Center}} = 6.65 - 8.78 = -2.13$$

b. Define the standard deviation to test normality:

$$s_{\text{Nonlinear}} = s_e\sqrt{\frac{1}{n} + \frac{1}{c}} = 0.615\sqrt{\frac{1}{12} + \frac{1}{5}} = 0.327$$

where

n = The total number of trials not at the center

c = The number of trials at the center.

c. Calculate the decision limits for nonlinearity.

$$\text{DL} = \pm (t_{\alpha/2,\text{df}}) \times s_{\text{Nonlinear}} = \pm 2.179 \times 0.327 = 0.71$$

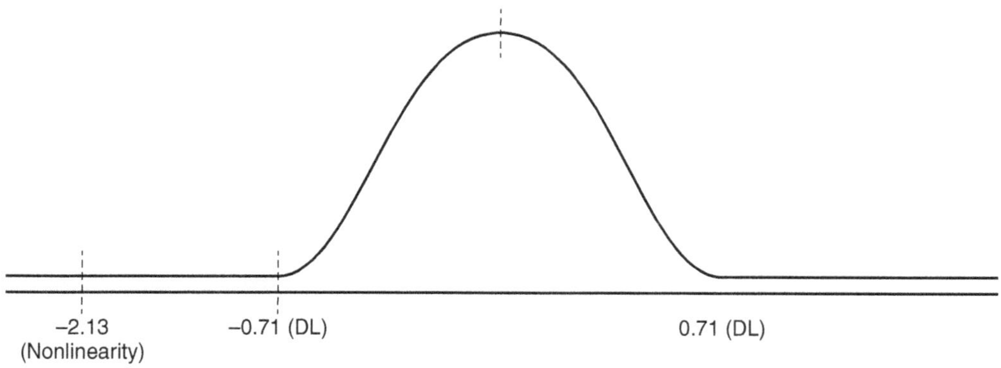

Figure 4.12 Decision limits for linearity.

With a typical alpha risk (also written with the Greek symbol α) of 5 percent, referring to Appendix C, Percentage Points of the Student's *t*-Distribution, for 12 degrees of freedom and $\alpha/2$, the critical value is $t = 2.179$.

Figure 4.12 illustrates that nonlinearity is significant. This means that the linear model cannot be used to optimize the process and that further defining of this process will require experimentation at three levels or more. It is safe to state only that the 0.10 percent nickel level is better than the 0.20 percent level, but we do not know what happens in between; we cannot interpolate. Since we do not know where the curvature is, we also cannot be sure of the conclusion that factor A (chromium) has no impact. We know only that these two specific levels are not significantly different in their responses.

At this point it would be wise to verify experimental results by using the model of the experiment to operate the process. All nonsignificant factors should be set solely on the basis of cost, productivity, or convenience. One may also consider opening up the window of the significant factors to see if further process improvement is possible.

5

Experiments with Three Factors

This chapter will move on to the next size DOE: one using three factors. The same analytical procedures are used. With three factors in a 2^3 factorial, there are eight runs, or combinations, of the three factors. Analysis columns for three main effects, three two-factor interactions, and one three-factor interaction can be computed from these eight basic runs. Although the three-factor interaction is discussed in this chapter, it is usually ignored since such interactions rarely occur.

5.1 CHEMICAL PROCESSING YIELD EXAMPLE

A team is striving to improve the yield of a chemical process. The controlling factors are hypothesized to be the processing temperature, the concentration level of a catalyst, and the *ramp time* (the time it takes to go from room temperature to the processing temperature). A formulation is processed for a specified time, cooled, and dissolved in a reagent. Then the end product is precipitated. The levels chosen for this study are:

A: Temperature—400 °F and 450 °F

B: Catalyst—10 percent and 20 percent

C: Ramp time—45 seconds and 90 seconds

The *response* is the yield of the refined product as a percentage of the raw stock. The objective is to maximize the yield. The design is a 2^3 factorial. Each run is replicated three times, with the 24 trials randomized as to order of test. The data and summary are shown in Table 5.1.

Nine Steps for Analysis of Effects

1. *Calculate the absolute values of the effects*:

$$E(A) = |\bar{Y}_{A+}| - |\bar{Y}_{A-}|$$

$$= \frac{71.77 + 70.58 + 68.27 + 73.93}{4} - \frac{62.16 + 55.42 + 61.70 + 56.40}{4} = 12.22$$

Table 5.1 Chemical processing yield data.

Run	A Temp	B Catalyst	C Ramp time	AB Interaction	AC Interaction	BC Interaction	ABC Interaction	Y^1	Y^2	Y^3	$\bar{Y}$
1	–	–	–	+	+	+	–	66.63	62.01	57.85	62.16
2	+	–	–	–	–	+	+	77.25	70.33	67.73	71.77
3	–	+	–	–	+	–	+	50.25	59.95	56.05	55.42
4	+	+	–	+	–	–	–	66.91	70.16	74.67	70.58
5	–	–	+	+	–	–	+	60.31	60.87	63.93	61.70
6	+	–	+	–	+	–	–	69.98	67.28	67.54	68.27
7	–	+	+	–	–	+	–	56.46	58.03	54.72	56.40
8	+	+	+	+	+	+	+	74.88	73.12	73.80	73.93
$\Sigma\bar{Y}+$	284.55	256.33	260.30	268.37	259.78	264.26	262.82				
$\Sigma\bar{Y}-$	235.68	263.90	259.93	251.86	260.45	255.97	257.41				$\bar{\bar{Y}}$ 65.03
$\bar{Y}+$	71.14	64.08	65.08	67.09	64.95	66.07	65.71				
$\bar{Y}-$	58.92	65.98	64.98	62.97	65.11	63.99	64.35				
C+	6.11	–0.95	0.05	2.06	–0.08	1.04	0.68				
C–	–6.11	0.95	–0.05	–2.06	0.08	–1.04	–0.68				
Abs effect	12.22	1.90	0.10	4.12	0.16	2.08	1.36				

$$E(B)=\left|\bar{Y}_{B+}\right|-\left|\bar{Y}_{B-}\right|$$

$$=\frac{55.42+70.58+56.40+73.93}{4}-\frac{62.16+71.77+61.70+68.27}{4}=1.90$$

$$E(C)=\left|\bar{Y}_{C+}\right|-\left|\bar{Y}_{C-}\right|$$

$$=\frac{61.70+68.27+56.40+73.93}{4}-\frac{62.16+71.77+55.42+70.58}{4}=0.10$$

$$E(AB)=\left|\bar{Y}_{AB+}\right|-\left|\bar{Y}_{AB-}\right|$$

$$=\frac{62.16+70.58+61.70+73.93}{4}-\frac{71.77+55.42+68.27+56.40}{4}=4.12$$

$$E(AC)=\left|\bar{Y}_{AC+}\right|-\left|\bar{Y}_{AC-}\right|$$

$$=\frac{62.16+55.42+68.27+73.93}{4}-\frac{71.77+70.58+61.70+56.40}{4}=0.16$$

$$E(BC)=\left|\bar{Y}_{BC+}\right|-\left|\bar{Y}_{BC-}\right|$$

$$=\frac{62.16+71.77+56.40+73.93}{4}-\frac{55.42+70.58+61.70+68.27}{4}=2.08$$

$$E(ABC)=\left|\bar{Y}_{ABC+}\right|-\left|\bar{Y}_{ABC-}\right|$$

$$=\frac{71.77+55.42+61.70+73.93}{4}-\frac{62.16+70.58+68.27+56.40}{4}=1.36$$

Note: The minus signs have been removed for simplicity.

2. *Create a Pareto chart of the absolute values of the effects.* The Pareto chart in Figure 5.1 indicates that factor A (temperature) and the interaction of AB (temperature and catalyst) may be significant.

3. *Calculate the standard deviation of the experiment, s_e:*

$$s_e=\sqrt{\left(\Sigma s^2\right)/k}$$

where

k = The number of runs

s^2 = The variance from each run

$$s_e=\sqrt{\left(\Sigma s^2\right)/k}=\sqrt{(19.27+24.21+23.81+15.21+3.80+2.22+2.76+0.79)/8}$$

$$=3.39$$

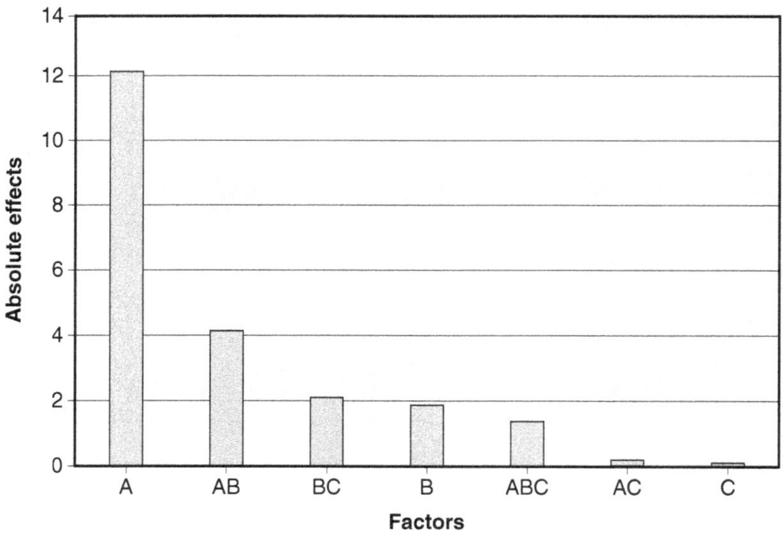

Figure 5.1 Pareto chart of the absolute values of the effects.

4. *Calculate the standard deviation of the effects,* s_{Eff}:

$$s_{\text{Eff}} = s_e\sqrt{4/n}$$

where

n = The total number of trials (Number of runs × Number of replicates)

s_e = The standard deviation

$$s_{\text{Eff}} = s_e\sqrt{4/n} = 3.39\sqrt{4/24} = 1.38.$$

5. *Determine the* t-*statistic*:

df = (# of observations per run – 1) × (# of runs) = (3 – 1) × (8) = 16.

Next, we must calculate a critical value to determine the significance of the effects. Confidence is 1 – alpha risk, where the alpha risk is the chance of erroneously claiming significance. The typical alpha risk (also written with the Greek symbol α) is therefore 5 percent. Referring to Appendix C, Percentage Points of the Student's t-Distribution, for 4 degrees of freedom and $\alpha/2$, the critical value is $t = 2.120$:

$$t_{\alpha/2,\text{df}} = t_{0.05/2,16 = 2.120}$$

6. *Calculate the decision limits*:

$$\text{DL} = (t_{\alpha/2,\text{df}})\,(s_{\text{Eff}}) = (2.120)\,(1.38) = 2.93$$

7. *Determine the significant effects.* An analysis of Figures 5.2 and 5.3 indicates that factor A (temperature) and the interaction of AB (temperature and catalyst) are significant.

8. *Graph the main effects and any significant interactions.* Figures 5.4, 5.5, 5.6, and 5.7 illustrate the effects and a significant interaction for the experiment.

9. *Model the significant effects, significant interactions, and individual effects from significant interactions.* Since our objective is to maximize yield, set factor A

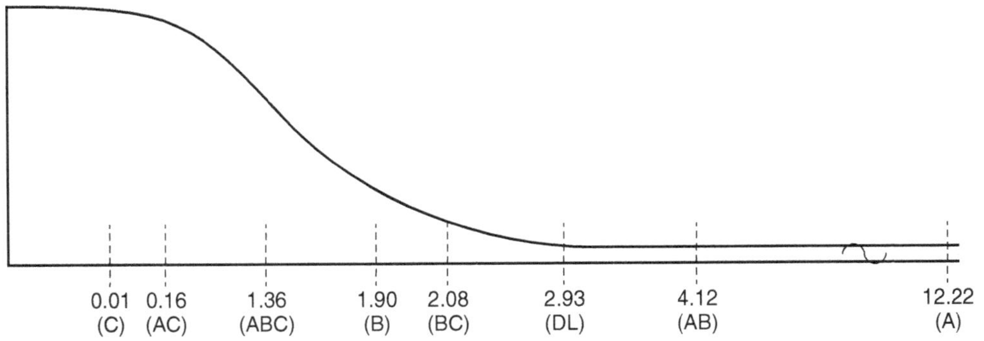

Figure 5.2 Decision limit for the effects.

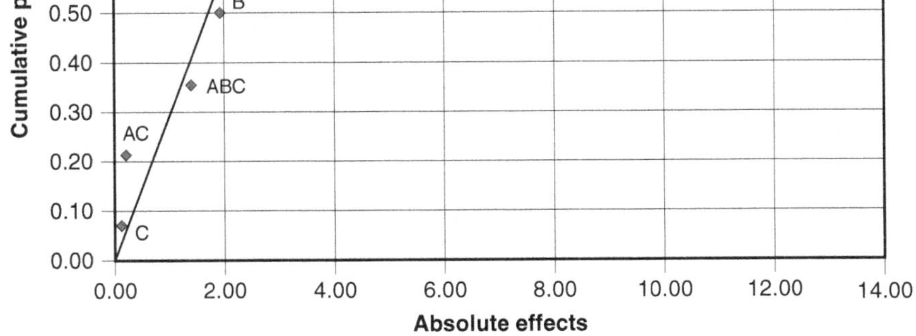

Figure 5.3 Half-normal plot of the absolute effects.

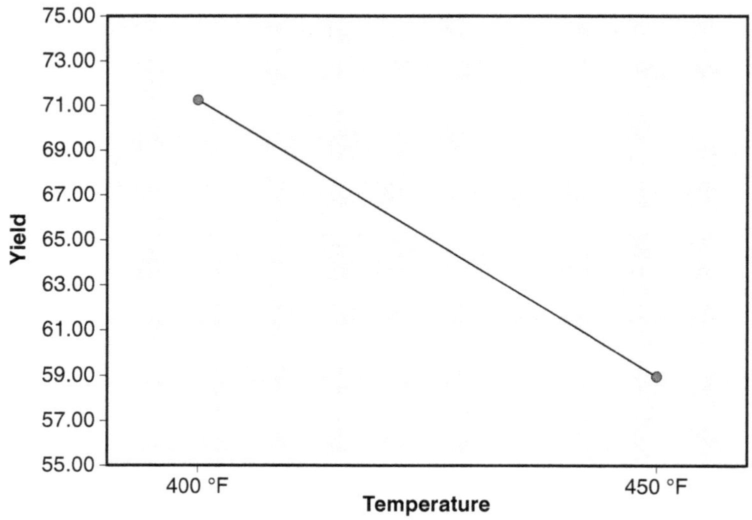

Figure 5.4 Plot of effect A (temperature) on yield.

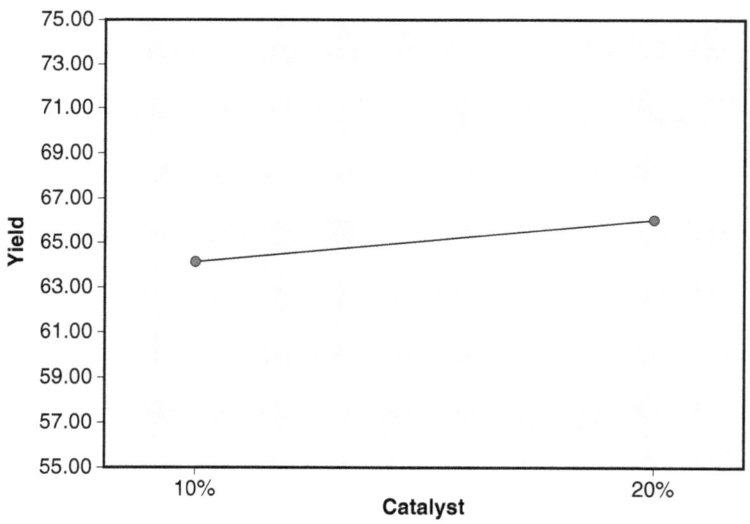

Figure 5.5 Plot of effect B (catalyst) on yield.

(temperature) to 450 °F and factor B (catalyst concentration) to 10% due to the interaction of AB. These levels will provide maximum yield because the interaction of AB is significant. Thus, to maximize the yield, Figure 5.7 indicates that factors A and B must be set to the (+), or high, levels due to the interaction. Factor C (ramp time, 45 seconds or 90 seconds) is not significant and can be set solely on the basis of cost, productivity, or convenience.

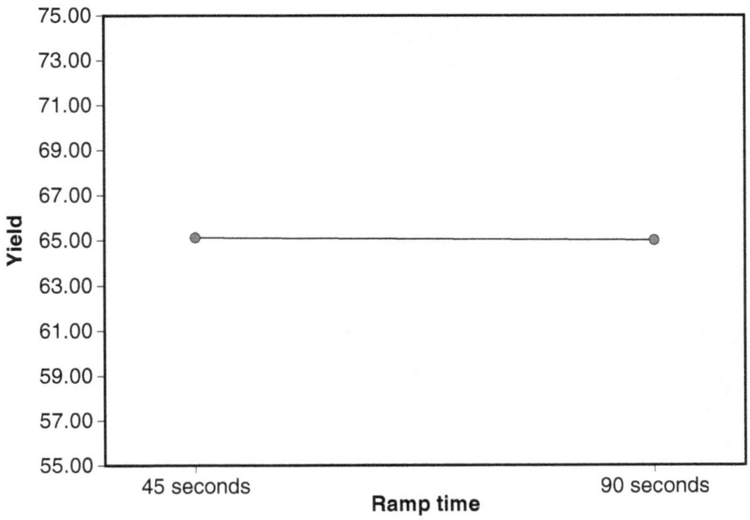

Figure 5.6 Plot of effect C (ramp time) on yield.

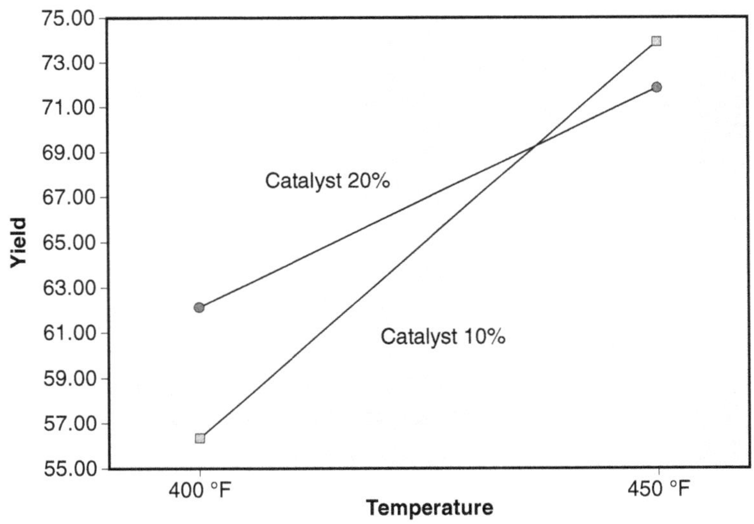

Figure 5.7 Plot of interaction AB (temperature-catalyst) on yield.

The coefficients must be calculated so the model can be compiled. The coefficient calculations are simply the average difference of the overall average from a level (+, –). Note that it is not necessary to calculate all of the coefficients. All coefficients are calculated for illustrative purposes.

$$C_{+,-} = \overline{Y}_{+,-} - \overline{\overline{Y}}$$

where

$\bar{\bar{Y}}$ = Average run S/N ratio

$\bar{Y}$ = Average at a treatment level.

$A_+ = 71.14 - 65.03 = 6.11$

$A_- = 58.92 - 65.03 = -6.11$

$B_+ = 64.08 - 65.03 = -0.95$

$B_-\ 65.98 - 65.03 = 0.95$

$C_+ = 65.08 - 65.03 = 0.05$

$C_- = 64.98 - 65.03 = -0.05$

$AB_+ = 67.09 - 65.03 = 2.06$

$AB_- = 62.97 - 65.03 = -2.06$

$AC_+ = 64.95 - 65.03 = -0.08$

$AC_- = 65.11 - 65.03 = 0.08$

$BC_+ = 66.07 - 65.03 = 1.04$

$BC_- = 63.99 - 65.03 = -1.04$

$ABC_+ = 65.71 - 65.03 = 0.68$

$ABC_- = 64.35 - 65.03 = -0.68$

Since our objective is to maximize yield, set factor A (temperature) to 450 °F and factor B (catalyst concentration) to 10% due to the interaction of AB). Refer to Figure 5.7.

$\hat{Y} = \bar{\bar{Y}}$ + Coefficients of the significant effects, significant interactions, and individual effects from significant interactions

$$\hat{Y} = \bar{\bar{Y}} + A_+ + B_+ + AB_+$$

$$\hat{Y} = 65.03 + 6.11 + -0.95 + 2.06 = 72.25$$

If the best combination of temperature and catalyst concentration is used, the yield is estimated to average 72.25. This is better than the results predicted simply from the main effects of A and B due to the contribution of the interaction term. An interaction either increases the response over that expected due to main effects alone, or it reduces the response below that expected from main effects.

At this point it would be wise to verify experimental results by using the model of the experiment to operate the process. All nonsignificant factors can be set solely on the basis of cost, productivity, or convenience. One may also consider opening up the window of the significant factors to see if further process improvement is possible. Remember that most problems are solved by a sequence of studies that move toward an optimum region. This further sequence of studies was developed by George E. P. Box in the 1950s and is referred to as *evolutionary operation* (EVOP). This technique further refines process variables in smaller increments to further optimize the desired process output.

5.2 VARIATION ANALYSIS

There is more relevant and useful information that can be utilized from the chemical processing yield experiment. What are the objectives of a DOE? They are to:

1. Control or move the process average
2. Reduce variation
3. Make the process more robust

The study of effects addresses both controlling or moving the process average and making the process more robust. However, reducing variation has not been addressed. A procedure is needed to determine whether any of the factors (or their interactions) will influence variation. This is done by analyzing the variances of the cells as the response for a variation analysis (see Table 5.2). The procedure for variation analysis is as follows:

1. Calculate the average variance for the high level, low level, and interaction of each factor and interaction:

$$S^2(A+) = \overline{S}^2_{A+} / n$$

$$S^2(A-) = \overline{S}^2_{A-} / n$$

$$S^2(B+) = \overline{S}^2_{B+} / n$$

$$S^2(B-) = \overline{S}^2_{B-} / n$$

$$S^2(C+) = \overline{S}^2_{C+} / n$$

$$S^2(C-) = \overline{S}^2_{C-} / n$$

$$S^2(AB+) = \overline{S}^2_{AB+} / n$$

$$S^2(AB-) = \overline{S}^2_{AB-} / n$$

$$S^2(AC+) = \overline{S}^2_{AC+} / n$$

$$S^2(AC-) = \overline{S}^2_{AC-} / n$$

Table 5.2 Chemical process yield example with variances.

Run	A Temp	B Catalyst	C Ramp time	AB Interaction	AC Interaction	BC Interaction	ABC Interaction	Y^1	Y^2	Y^3	$\bar{Y}$	s^2
1	–	–	–	+	+	+	–	66.63	62.01	57.85	62.16	19.27
2	+	–	–	–	–	+	+	77.25	70.33	67.73	71.77	24.21
3	–	+	–	–	+	–	+	50.25	59.95	56.05	55.42	23.81
4	+	+	–	+	–	–	–	66.91	70.16	74.67	70.58	15.21
5	–	–	+	+	–	–	+	60.31	60.87	63.93	61.70	3.80
6	+	–	+	–	+	–	–	69.98	67.28	67.54	68.27	2.22
7	–	+	+	–	–	+	–	56.46	58.03	54.72	56.40	2.76
8	+	+	+	+	+	+	+	74.88	73.12	73.80	73.93	0.79
$\Sigma\bar{Y}+$	284.55	256.33	260.30	268.37	259.78	264.26	262.82					
$\Sigma\bar{Y}-$	235.68	263.90	259.93	251.86	260.45	255.97	257.41				$\bar{\bar{Y}}$ 65.03	
$\bar{Y}+$	71.14	64.08	65.08	67.09	64.95	66.07	65.71					
$\bar{Y}-$	58.92	65.98	64.98	62.97	65.11	63.99	64.35					
C+	6.11	–0.95	0.05	2.06	–0.08	1.04	0.68					
C–	–6.11	0.95	–0.05	–2.06	0.08	–1.04	–0.68					
Abs effect	12.22	1.90	0.10	4.12	0.16	2.08	1.36					
$\bar{s}^2_+$	10.61	10.64	2.39	9.77	11.52	11.76	13.15					
$\bar{s}^2_-$	12.41	12.38	20.36	13.25	11.50	11.26	9.87					
F	1.17	1.16	8.63	1.36	1.00	1.04	1.33					

$$S^2(BC+) = \overline{S}^2_{BC+} / n$$

$$S^2(BC-) = \overline{S}^2_{BC-} / n$$

$$S^2(ABC+) = \overline{S}^2_{ABC+} / n$$

$$S^2(ABC-) = \overline{S}^2_{ABC-} / n$$

where

n = The number of runs/2.

For the chemical processing yield example (see Table 5.2):

$$S^2(A+) = \overline{S}^2_{A+} / n = \frac{24.21 + 15.21 + 2.22 + 0.79}{4} = 10.61$$

$$S^2(A-) = \overline{S}^2_{A-} / n = \frac{19.27 + 23.81 + 3.80 + 2.76}{4} = 12.41$$

$$S^2(B+) = \overline{S}^2_{B+} / n = \frac{23.81 + 15.21 + 2.76 + 0.79}{4} = 10.64$$

$$S^2(B-) = \overline{S}^2_{B-} / n = \frac{19.27 + 24.21 + 3.80 + 2.22}{4} = 12.38$$

$$S^2(C+) = \overline{S}^2_{C+} / n = \frac{3.80 + 2.22 + 2.76 + 0.79}{4} = 2.39$$

$$S^2(C-) = \overline{S}^2_{C-} / n = \frac{19.27 + 24.21 + 23.81 + 15.21}{4} = 20.63$$

$$S^2(AB+) = \overline{S}^2_{AB+} / n = \frac{19.27 + 15.21 + 3.80 + 0.79}{4} = 9.77$$

$$S^2(AB-) = \overline{S}^2_{AB-} / n = \frac{24.21 + 23.81 + 2.22 + 2.76}{4} = 13.25$$

$$S^2(AC+) = \overline{S}^2_{AC+} / n = \frac{19.27 + 23.81 + 2.22 + 0.79}{4} = 11.52$$

$$S^2(AC-) = \overline{S}^2_{AC-} / n = \frac{24.21 + 15.21 + 3.80 + 2.76}{4} = 11.50$$

$$S^2(BC+) = \overline{S}^2_{BC+} / n = \frac{19.27 + 24.21 + 2.76 + 0.79}{4} = 11.76$$

$$S^2(BC-) = \overline{S}^2_{BC-} / n = \frac{23.81 + 15.21 + 3.80 + 2.22}{4} = 11.26$$

$$S^2(ABC+) = \overline{S}^2_{ABC+} / n = \frac{24.21 + 23.81 + 3.80 + 0.79}{4} = 13.15$$

$$S^2(ABC-) = \overline{S}^2_{ABC-} / n = \frac{19.29 + 15.21 + 2.22 + 2.76}{4} = 9.87$$

2. *For each factor and interaction, calculate a new statistic*: $F = S^2$ larger/S^2 smaller. For the chemical processing yield example (see Table 5.2):

$$F(A) = \frac{12.41}{10.61} = 1.17$$

$$F(B) = \frac{12.38}{10.64} = 1.16$$

$$F(C) = \frac{20.63}{2.39} = 8.63$$

$$F(AB) = \frac{13.25}{9.77} = 1.36$$

$$F(AC) = \frac{11.52}{11.50} = 1.00$$

$$F(BC) = \frac{11.76}{11.26} = 1.04$$

$$F(ABC) = \frac{13.15}{9.87} = 1.33$$

3. *Determine* F_{Critical}:

df = ((# of observations per run – 1) × (# of runs))/2 = ((3 – 1) × (8))/2 = 8.

F_{Critical} = α, df, df values of the *F*-distribution (See Appendix B, Percentages of the *F*-Distribution).

For this example we will use an α of 0.05.

$$F_{(\alpha 0.05, df8, df8)} = 3.44$$

4. *Plot the F-statistic for each factor and interaction.* Only one factor (C: ramp time) has an *F*-value exceeding that test statistic (see Figure 5.8). Since the *F* is based on the larger variance divided by the smaller, one must check the individual variances to see that the high (+) ramp time level will provide lower variation.

5. *Evaluate the conclusions for variation. Determine optimum operating conditions for both effects and variation. If the two objectives conflict, a compromise must be reached.* The experimenter's conclusions must now be modified. Based on effects alone,

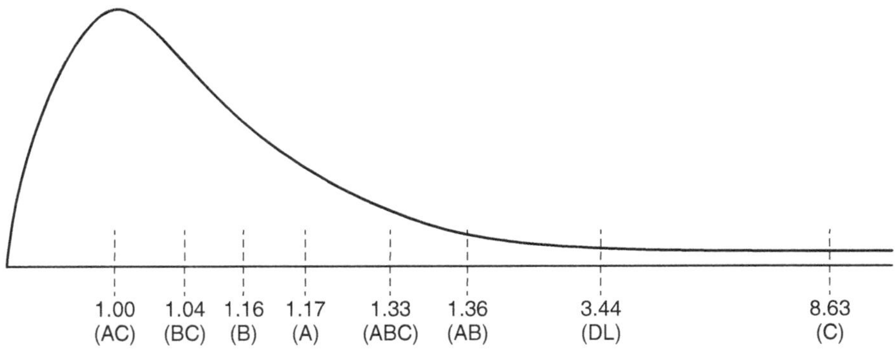

Figure 5.8 Decision limit for variances.

there was no reason to be concerned about the level of ramp time because it was not significant. Adding the variation analysis, it is now apparent that it is quite important to have a longer ramp time in order to reduce the variation in the yield. These conclusions have a major impact on the estimate of inherent variation for the predictions from the model. If ramp time is at the higher (+) level, the inherent variation under those conditions is estimated by

$$s_e = \sqrt{2.39} = 1.55.$$

If the lower level (–) of ramp time is used, the responses have an estimate of inherent variation of

$$s_e = \sqrt{20.63} = 4.54.$$

Estimation of process capability is impacted by which standard deviation is used. More importantly, the inherent variation in the process is affected. Some general guidelines for this joint analysis of effects and variation follow:

1. Significance of effects is independent of the assessment of variation, and vice versa. They are two uncorrelated assessments.

2. It is possible to be unlucky and have a conflict; that is, a factor may occasionally be significant for both effects and variation, with different levels required for the optimization of each. This requires a compromise or trade-off in trying to achieve an adequate, if not ideal, operating condition for both objectives.

3. Variation analysis is particularly susceptible to outliers or special cause variation that results from influences outside the DOE. These occurrences distort the assessment of variation. An indicator of this problem is the occurrence of more than one significant term in the variation analysis.

4. If an effect that will control process averages is worth its weight in gold (a *golden* factor), then a factor that controls variation is a *diamond* factor. Such factors are relatively rare, and, when they occur, they may be the most important discovery in an experiment. Generally speaking, factors can be found to change the averages whereas it is difficult to find a factor that reduces variation.

When an experiment is replicated (two or more trials for each run), the assessment of variation is a free extension of the analysis since the data are already available without additional testing required.

5.3 ANALYSIS WITH UNREPLICATED EXPERIMENTS (RESIDUAL ANALYSIS)

What can be done for analysis if there is no replication? This is a common occurrence in DOE due to the expense of the experimentation. The usual procedures will not work since there are no variances without replication. Consider the abbreviated analysis table in Table 5.3 for an unreplicated experiment with three factors (2^3). Following is a variation of a technique discussed in Montgomery (2013) that can be applied even though there are no variances to compare. We will use the chemical processing yield example to demonstrate the residual procedure.

1. *Calculate the predicted response for each run by using the model and enter it in the column for predictions.* The model contains the significant effects, significant interactions, and individual effects from significant interactions (see Table 5.3). The experimental model was determined to be:

$$\hat{Y} = \bar{\bar{Y}} + \text{Coefficients of the significant effects, significant interactions, and individual effects from significant interactions}$$

$$\hat{Y} = \bar{\bar{Y}} + A_+ + B_+ + AB_+$$

$$\hat{Y} = 65.03 + 6.11 + -0.95 + 2.06 = 72.25$$

However, for residual analysis, a model must be constructed for each run using

$$\hat{Y}(1) = \bar{\bar{Y}} + A_- + B_- + AB_+ = 65.03 + -6.11 + 0.95 + 2.06 = 61.93$$

$$\hat{Y}(2) = \bar{\bar{Y}} + A_+ + B_- + AB_+ = 65.03 + 6.11 + 0.95 + -2.06 = 70.03$$

$$\hat{Y}(3) = \bar{\bar{Y}} + A_- + B_+ + AB_- = 65.03 + -6.11 + -0.95 + -2.06 = 55.91$$

$$\hat{Y}(4) = \bar{\bar{Y}} + A_+ + B_+ + AB_+ = 65.03 + 6.11 + -0.95 + 2.06 = 72.25$$

Table 5.3 Chemical processing yield data (unreplicated).

Run	A Temp	B Catalyst	C Ramp time	AB Interaction	AC Interaction	BC Interaction	ABC Interaction	*Y*	Pred	Res
1	–	–	–	+	+	+	–	62.16	61.93	0.23
2	+	–	–	–	–	+	+	71.77	70.03	1.74
3	–	+	–	–	+	–	+	55.42	55.91	–0.49
4	+	+	–	+	–	–	–	70.58	72.25	–1.67
5	–	–	+	+	–	–	+	61.70	61.93	–0.23
6	+	–	+	–	+	–	–	68.27	70.03	–1.76
7	–	+	+	–	–	+	–	56.40	55.91	0.49
8	+	+	+	+	+	+	+	73.93	72.25	1.68
s_+^2	1.98	0.98	1.44	1.38	1.44	0.79	1.20			
s_-^2	0.44	1.44	1.43	1.48	1.42	0.79	1.20			
F	4.50	1.47	1.01	1.07	1.01	1.00	1.00			

$$\hat{Y}(5)=\overline{\overline{Y}}+A_{-}+B_{-}+AB_{+}=65.03+-6.11+0.95+2.06=61.93$$

$$\hat{Y}(6)=\overline{\overline{Y}}+A_{+}+B_{-}+AB_{-}=65.03+6.11+0.95+-2.06=70.03$$

$$\hat{Y}(7)=\overline{\overline{Y}}+A_{-}+B_{+}+AB_{-}=65.03+-6.11+-0.95+-2.06=55.91$$

$$\hat{Y}(8)=\overline{\overline{Y}}+A_{+}+B_{+}+AB_{+}=65.03+6.11+0.95+2.06=72.25$$

2. *Calculate a residual for each run by*

Response value for each run – Predicted value for each run

Run(1) 62.16 – 61.93 = 0.23

Run(2) 71.77 – 70.03 = 1.74

Run(3) 55.42 – 55.91 = –0.49

Run(4) 70.58 – 72.25 = –1.67

Run(5) 61.70 – 61.93 = –0.23

Run(6) 68.27 – 70.03 = –1.76

Run(7) 56.40 – 55.91 = 0.49

Run(8) 73.93 – 72.25 = 1.68

4. *Check the residuals for normality using a normal probability plot.* The normal plot of residuals (Figure 5.9) indicates that the residuals are normally distributed. When there is an extreme departure from normality, further analysis is necessary. This is especially true when an outlier is present that can adversely affect the variance.

5. *For each column, calculate the variance of the residuals for the high (+) level of the column and then for the low (–) level of the column.*

6. *For each factor and interaction, calculate a new statistic*: $F = s^2$ larger/s^2 smaller. For the chemical processing yield example (see Table 5.3):

$$F(A)=\frac{1.98}{0.44}=4.50$$

$$F(B)=\frac{1.44}{0.98}=1.47$$

$$F(C)=\frac{1.44}{1.43}=1.01$$

$$F(AB)=\frac{1.48}{1.38}=1.07$$

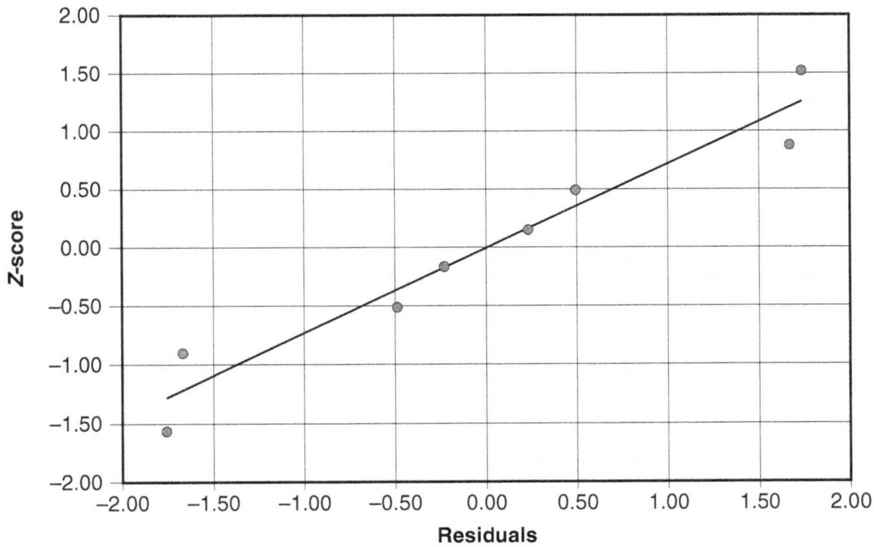

Figure 5.9 Normal plot of residuals.

$$F(AC)=\frac{1.44}{1.42}=1.01$$

$$F(BC)=\frac{0.79}{0.79}=1.00$$

$$F(ABC)=\frac{1.20}{1.20}=1.00$$

7. *Determine* F_{Critical}: $F_{\text{Critical}} = \alpha$, df, df values of the F-distribution (see Appendix B, Percentages of the F-Distribution). The degrees of freedom are the number of residuals used to calculate the standard deviation –1. In this case the df = 3.

For this example we will use an α of 0.05.

$$F_{(\alpha 0.05,\text{df}3,\text{df}3)}=9.28$$

Since factor A: temperature, exceeds the test value f from the F-tests performed (see Figure 5.10), factor A: temperature is significant with respect to its influence on variation. Recall from Section 5.2, Variation Analysis, that one factor (C: ramp time) at the high (+) ramp time level will provide lower variation. The difference in these approaches is due to the residual analysis technique being weak (it will only detect very large differences in variances) when the total degrees of freedom are less than 14. With larger experiments, it is a very useful tool.

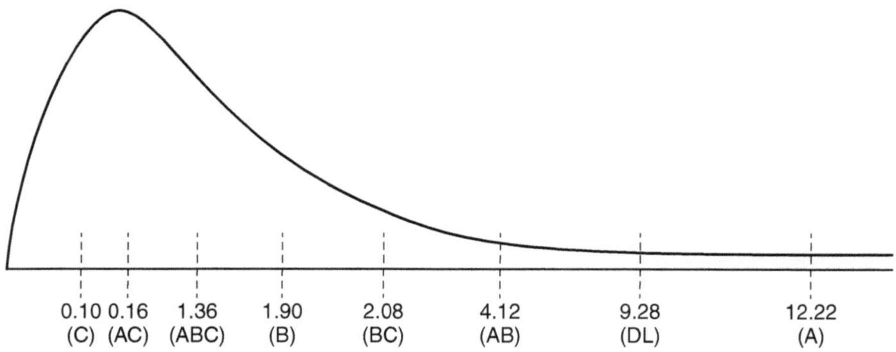

Figure 5.10 Decision limit for variances.

While this approach looks tedious, remember that it is actually free information; that is, it requires no additional data. This technique can be easily applied with a spreadsheet and can be automatically calculated with specialized software.

6

Experiments with Qualitative (Attribute Data) Responses

There are times when variable output responses are not practical when conducting a DOE. In these cases qualitative data must be used (go/no-go, pass/fail, or categorical). Visual inspections are a good example of qualitative responses. Using attribute responses can also be very expensive due to the number of replicates potentially necessary to detect and collect failures. For example, a process with a 1% defect rate means that generally there is one defect for every 100 parts produced. In this case several hundred parts will need to be produced for each run to "ensure" that defects can be detected. In order to overcome using large sample sizes, it may become necessary to open or intensify (Ross 1996) the factor levels (increase the upper level and decrease the lower level) to better understand the impact of the factors being studied. This method will effectively "force" the results and help determine the impact of the process.

The importance of a good measurement systems analysis (MSA), test method validation (TMV), or gage repeatability and reproducibility (GR&R) study cannot be understated for qualitative output responses, and must be considered a prerequisite to executing a DOE.

When the data are not normal, as in the case of percentages and probabilities, a data transformation is necessary to normalize the data to allow the use of parametric statistics.

6.1 PLASTIC WELDING EXAMPLE (WITHOUT TRANSFORMATION)

A plastic weld process has been generating a significant amount of defective product. The defect cannot be measured using quantitative (variable) data and is assessed using pass/fail criteria. A GRR was performed to ensure the adequacy of the measuring system. The measuring system was found to be adequate for its intended purpose. The goal of the experiment is to eliminate the defect or reduce the occurrence. The process has three key variables: time, temperature, and pressure. The team decided, based on experience and the manufacturer's recommendation, to use the following levels:

A: Time—1.25 seconds and 2.50 seconds

B: Temperature—250 °C and 260 °C

C: Pressure—80 psi and 90 psi

100 samples were produced for each run. The results are shown in Table 6.1. The output response (Y) is the count of the number of defects detected.

Nine Steps for Analysis of Effects

1. *Calculate the absolute values of the effects*:

$$|E(A)| = |\bar{Y}_{A+}| - |\bar{Y}_{A-}| = \frac{5.00+0.00+3.00+0.00}{4} - \frac{8.00+4.00+9.00+5.00}{4} = 4.50$$

$$|E(B)| = |\bar{Y}_{B+}| - |\bar{Y}_{B-}| = \frac{4.00+0.00+5.00+0.00}{4} - \frac{8.00+5.00+9.00+3.00}{4} = 4.00$$

$$|E(C)| = |\bar{Y}_{C+}| - |\bar{Y}_{C-}| = \frac{8.00+5.00+4.00+0.00}{4} - \frac{9.00+3.00+5.00+0.00}{4} = 0.00$$

$$|E(AB)| = |\bar{Y}_{AB+}| - |\bar{Y}_{AB-}| = \frac{8.00+0.00+9.00+0.00}{4} - \frac{5.00+4.00+3.00+5.00}{4}$$

$$= 0.00$$

$$|E(AC)| = |\bar{Y}_{AC+}| - |\bar{Y}_{AC-}| = \frac{8.00+4.00+3.00+0.00}{4} - \frac{5.00+0.00+9.00+5.00}{4}$$

$$= 1.00$$

$$|E(BC)| = |\bar{Y}_{BC+}| - |\bar{Y}_{BC-}| = \frac{8.00+5.00+5.00+0.00}{4} - \frac{4.00+0.00+9.00+3.00}{4}$$

$$= 0.50$$

$$|E(ABC)| = |\bar{Y}_{ABC+}| - |\bar{Y}_{ABC-}| = \frac{5.00+4.00+9.00+0.00}{4} - \frac{8.00+0.00+3.00+5.00}{4}$$

$$= 0.50$$

Note: The minus signs have been removed for simplicity.

2. *Create a Pareto chart of the absolute values of the effects.* The Pareto chart in Figure 6.1 indicates that factors A (time) and B (temperature) are significant. The interaction of AC (time and pressure) may also be significant.

3. *Calculate the standard deviation of the experiment.* This step does not apply with qualitative data.

4. *Calculate the standard deviation of the effects.* This step does not apply with qualitative data.

5. *Determine the statistic.* This step does not apply with qualitative data.

Table 6.1 Plastic welding without transformation example data.

Run	A Time	B Temp	C Pressure	AB Interaction	AC Interaction	BC Interaction	ABC Interaction	*Y*
1	–	–	–	+	+	+	–	8
2	+	–	–	–	–	+	+	5
3	–	+	–	–	+	–	+	4
4	+	+	–	+	–	–	–	0
5	–	–	+	+	–	–	+	9
6	+	–	+	–	+	–	–	3
7	–	+	+	–	–	+	–	5
8	+	+	+	+	+	+	+	0
$\Sigma\bar{Y}+$	8.00	9.00	17.00	17.00	15.00	18.00	18.00	
$\Sigma\bar{Y}-$	26.00	25.00	17.00	17.00	19.00	16.00	16.00	$\bar{\bar{Y}}$ 4.25
$\bar{Y}+$	2.00	2.25	4.25	4.25	3.75	4.50	4.50	
$\bar{Y}-$	6.50	6.25	4.25	4.25	4.75	4.00	4.00	
C+	–2.25	–2.00	0.00	0.00	–0.50	0.25	0.25	
C–	2.25	2.00	0.00	0.00	0.50	–0.25	–0.25	
Abs effect	4.50	4.00	0.00	0.00	1.00	0.50	0.50	

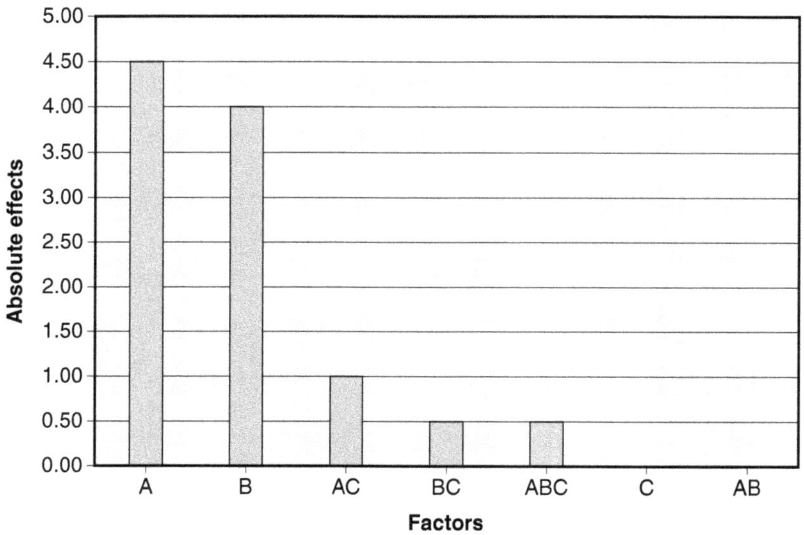

Figure 6.1 Pareto chart of the absolute values of the effects.

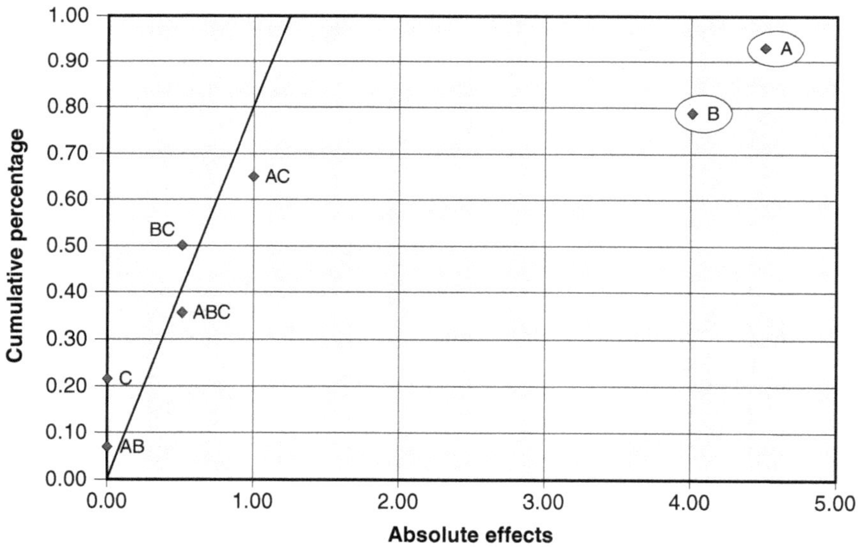

Figure 6.2 Half-normal plot of the absolute effects.

6. *Calculate the decision limits.* This step will be done with the use of a half-normal plot (Figure 6.2).

7. *Determine the significant effects.* An analysis of Figure 6.2 indicates that factors A (time) and B (temperature) are significant.

8. *Graph the main effects and any significant interactions* (Figures 6.3, 6.4, and 6.5).

9. *Model the significant effects, significant interactions, and individual effects from significant interactions.* Since our objective is to minimize defects, factor A (time 1.25 or 2.50 seconds) and factor B (temperature 250 °C or 260 °C) are set at the high (+1) levels. These levels will provide the lowest number of defects. Factor C (pressure 80 psi

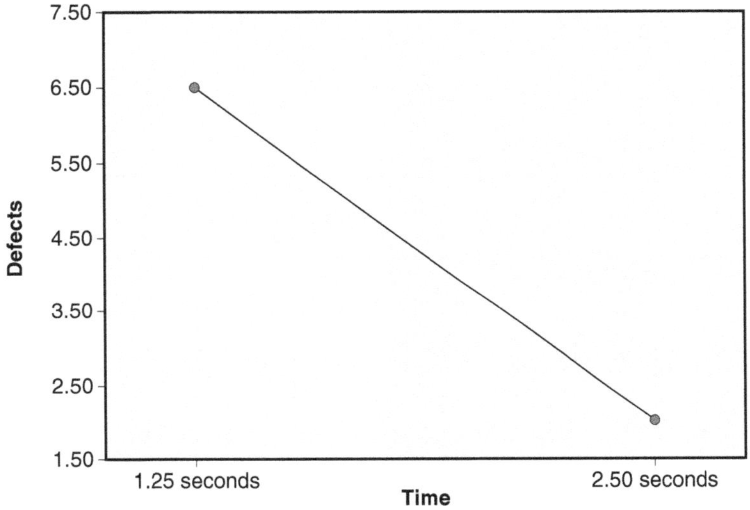

Figure 6.3 Effects plot for effect A (time).

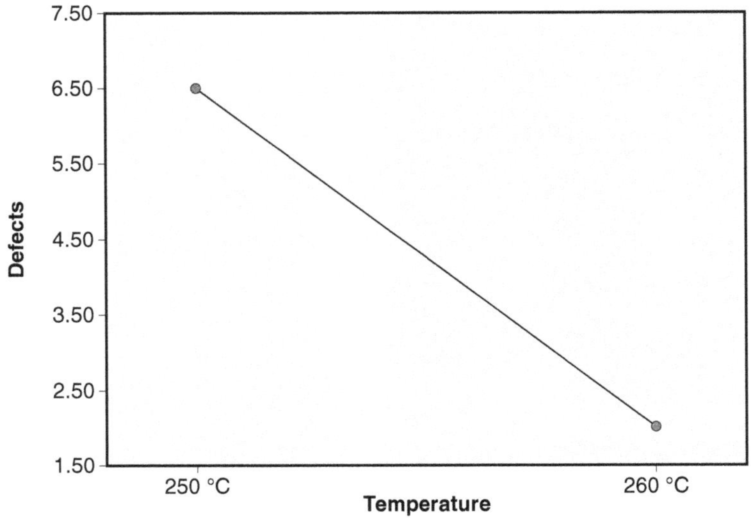

Figure 6.4 Effects plot for effect B (temperature).

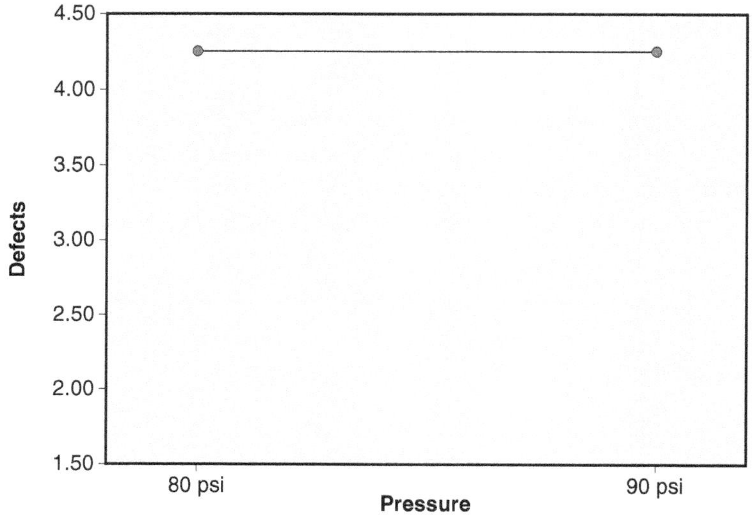

Figure 6.5 Effects plot for effect C (pressure).

or 90 psi) is not significant and can be set solely on the basis of cost, productivity, or convenience.

The coefficients must be calculated so the model can be compiled. The coefficient calculations are simply the average difference of the overall average from a level (+, –). Note that it is not necessary to calculate all of the coefficients. All coefficients are calculated for illustrative purposes.

$$C_{+,-} = \bar{Y}_{+,-} - \bar{\bar{Y}}$$

where

$\bar{\bar{Y}}$ = Average run S/N ratio

$\bar{Y}$ = Average at a treatment level

$A_+ = 2.00 - 4.25 = -2.25$

$A_- = 6.50 - 4.25 = 2.25$

$B_+ = 2.25 - 4.25 = -2.00$

$B_- = 6.25 - 4.25 = 2.00$

$C_+ = 4.25 - 4.25 = 0.00$

$C_- = 4.25 - 4.25 = 0.00$

$AB_+ = 4.25 - 4.25 = 0.00$

$AB_- = 4.25 - 4.25 = 0.00$

$AC_+ = 3.75 - 4.25 = -0.50$

$$AC_- = 4.75 - 4.25 = 0.50$$

$$BC_+ = 4.50 - 4.25 = 0.25$$

$$BC_- = 4.00 - 4.25 = -0.25$$

$$ABC_+ = 4.50 - 4.25 = 0.25$$

$$ABC_- = 4.00 - 4.25 = -0.25$$

Since our objective is to minimize defects, set factor A (time) to 2.50 seconds and factor B (temperature) to 260 °C. Refer to Figures 6.3 and 6.4.

$\hat{Y} = \overline{\overline{Y}} +$ Coefficients of the significant effects, significant interactions, and individual effects from significant interactions

$$\hat{Y} = \overline{\overline{Y}} + A_+ + B_+$$

$$\hat{Y} = 4.25 + -2.25 + -2.00 = 0$$

At this point it would be wise to verify experimental results by using the model of the experiment to operate the process. All nonsignificant factors can be set solely on the basis of cost, productivity, or convenience. One may also consider opening up the window of the significant factors to see if further process improvement is possible. Remember that most problems are solved by a sequence of studies that move toward an optimum region. This requires that one be aware of physical and safety constraints that may limit further studies. If still not satisfied with the performance, look for other factors.

This procedure has been effectively used on several occasions with great success. However, from a statistical point of view, several shortcuts were liberally used. When the data transformation is applied, the results are substantially the same with a few minor exceptions.

6.2 PLASTIC WELDING EXAMPLE (WITH TRANSFORMATION)

We will now perform a data transformation to normalize the data from Section 6.1. Recall that the process has three key variables: time, temperature, and pressure. The team decided, based on experience and the manufacturer's recommendation, to use the following levels:

A: Time—1.25 seconds and 2.50 seconds

B: Temperature—250 °C and 260 °C

C: Pressure—80 psi and 90 psi

To properly normalize the data, the number of defects in the response must be converted to a percentage. 100 samples were produced for each run. The results are shown in Table

6.2. The output response (Y) is the count of the number of defects detected. In this case it is a simple matter of dividing the response of each run by 100. The next step is to take the square root of the percentage for each run. The final step in transforming the data is to take the arcsine or $\sin^{-1}$ of the square root of the percentage for each run (see Table 6.2). The arcsine transformation will work using either degrees or radians. All percentage values must be from 0.00 to 1.00 inclusive. This transformation is not appropriate for percentages greater than 100% as in the case of a percentage increase.

Nine Steps for Analysis of Effects

1. *Calculate the absolute values of the effects:*

$$|E(A)| = |\bar{Y}_{A+}| - |\bar{Y}_{A-}|$$

$$= \frac{12.70+0.00+9.79+0.00}{4} - \frac{16.26+11.54+17.47+12.71}{4} = 8.87$$

$$|E(B)| = |\bar{Y}_{B+}| - |\bar{Y}_{B-}|$$

$$= \frac{11.54+0.00+12.71+0.00}{4} - \frac{16.26+12.70+17.47+9.79}{4} = 7.99$$

$$|E(C)| = |\bar{Y}_{C+}| - |\bar{Y}_{C-}|$$

$$= \frac{16.26+12.70+11.54+0.00}{4} - \frac{17.47+9.79+12.71+0.00}{4} = 0.14$$

$$|E(AB)| = |\bar{Y}_{AB+}| - |\bar{Y}_{AB-}|$$

$$= \frac{16.26+0.00+17.47+0.00}{4} - \frac{12.70+11.54+9.79+12.71}{4} = 3.26$$

$$|E(AC)| = |\bar{Y}_{AC+}| - |\bar{Y}_{AC-}|$$

$$= \frac{16.26+11.54+9.79+0.00}{4} - \frac{12.70+0.00+17.47+12.71}{4} = 1.32$$

$$|E(BC)| = |\bar{Y}_{BC+}| - |\bar{Y}_{BC-}|$$

$$= \frac{16.26+12.70+12.71+0.00}{4} - \frac{11.54+0.00+17.47+9.79}{4} = 0.72$$

$$|E(ABC)| = |\bar{Y}_{ABC+}| - |\bar{Y}_{ABC-}|$$

$$= \frac{12.70+11.54+17.47+0.00}{4} - \frac{16.26+0.00+9.79+12.71}{4} = 0.74$$

Table 6.2 Plastic welding with transformation example data.

Run	A Time	B Temp	C Pressure	AB Interaction	AC Interaction	BC Interaction	ABC Interaction	Y	%	SQRT	$\bar{Y}$ ARCSIN
1	–	–	–	+	+	+	–	8.00	0.08	0.28	16.26
2	+	–	–	–	–	+	+	5.00	0.05	0.22	12.70
3	–	+	–	–	+	–	+	4.00	0.04	0.20	11.54
4	+	+	–	+	–	–	–	0.00	0.00	0.00	0.00
5	–	–	+	+	–	–	+	9.00	0.09	0.30	17.47
6	+	–	+	–	+	–	–	3.00	0.03	0.17	9.79
7	–	+	+	–	–	+	–	5.00	0.05	0.22	12.71
8	+	+	+	+	+	+	+	0.00	0.00	0.00	0.00
$\Sigma\bar{Y}+$	22.49	24.25	40.50	33.73	37.59	41.67	41.71				
$\Sigma\bar{Y}-$	57.98	56.22	39.97	46.74	42.88	38.80	38.76				$\bar{\bar{Y}}$ 10.06
$\bar{Y}+$	5.62	6.06	10.13	8.43	9.40	10.42	10.43				
$\bar{Y}-$	14.50	14.06	9.99	11.69	10.72	9.70	9.69				
C+	–4.44	–4.00	0.06	–1.63	–0.66	0.36	0.37				
C–	4.44	4.00	–0.07	1.63	0.66	–0.36	–0.37				
Abs effect	8.87	7.99	0.14	3.26	1.32	0.72	0.74				

Note: the minus signs have been removed for simplicity.

2. *Create a Pareto chart of the absolute values of the effects.* The Pareto chart in Figure 6.6 indicates that factors A (time) and B (temperature) are significant. The interaction of AB (time and temperature) may also be significant.

3. *Calculate the standard deviation of the experiment.* This step does not apply with qualitative data.

4. *Calculate the standard deviation of the effects.* This step does not apply with qualitative data.

5. *Determine the statistic.* This step does not apply with qualitative data.

6. *Calculate the decision limits.* This step will be done with the use of a half-normal plot (Figure 6.7).

7. *Determine the significant effects.* An analysis of Figure 6.7 indicates that factors A (time), B (temperature), and the interaction AB (time and temperature) are significant.

8. *Graph the main effects and any significant interactions* (Figures 6.8, 6.9, 6.10, and 6.11).

9. *Model the significant effects, significant interactions, and individual effects from significant interactions.* Since our objective is to minimize defects, factor A (time) should be set to 2.50 seconds, and factor B (temperature) should be set to 260 °C, the high (+1) levels. These levels will provide minimum defects because the interaction of AB is significant. Thus, to minimize defects, Figure 6.11 indicates that factors A and B must be set to the (+), or high, levels due to the interaction of AB. These levels will

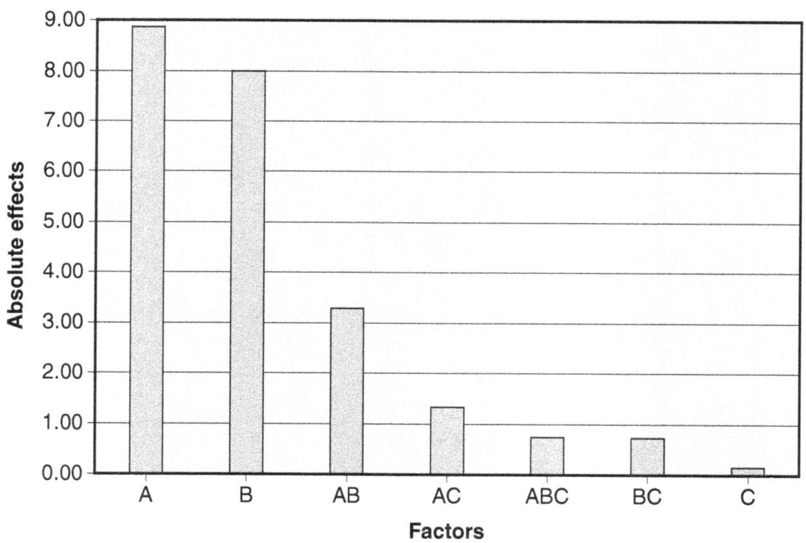

Figure 6.6 Pareto chart of the absolute values of the effects.

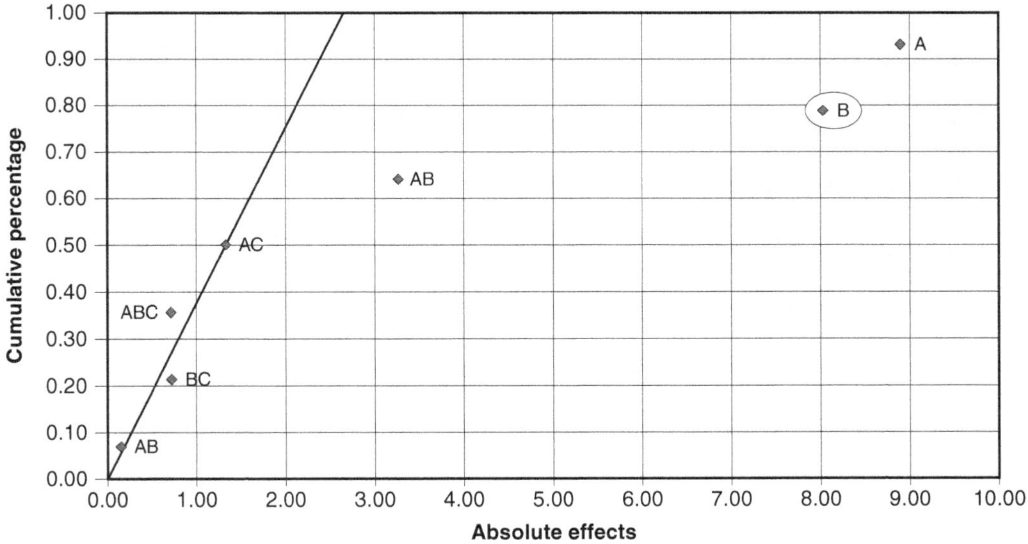

Figure 6.7 Half-normal plot of the absolute effects.

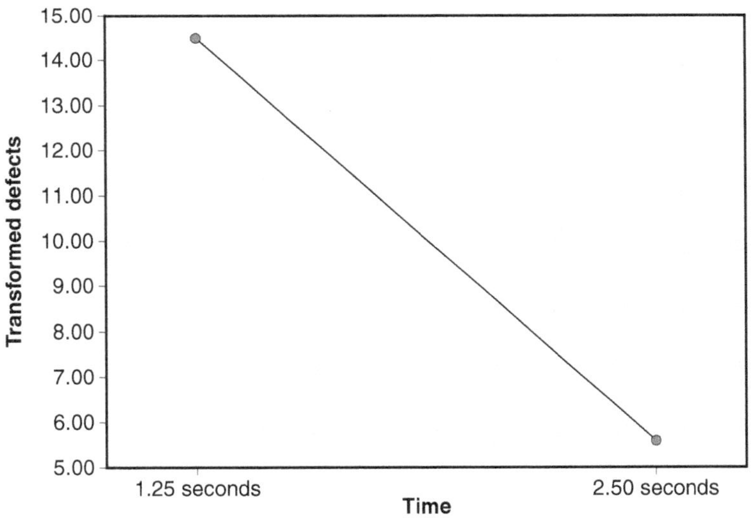

Figure 6.8 Plot of effect A (time) on transformed defects.

provide the lowest number of defects. Factor C (pressure 80 psi or 90 psi) is not significant and can be set solely on the basis of cost, productivity, or convenience.

The coefficients must be calculated so the model can be compiled. The coefficient calculations are simply the average difference of the overall average from a level (+, –). Note that it is not necessary to calculate all of the coefficients. All coefficients are calculated for illustrative purposes.

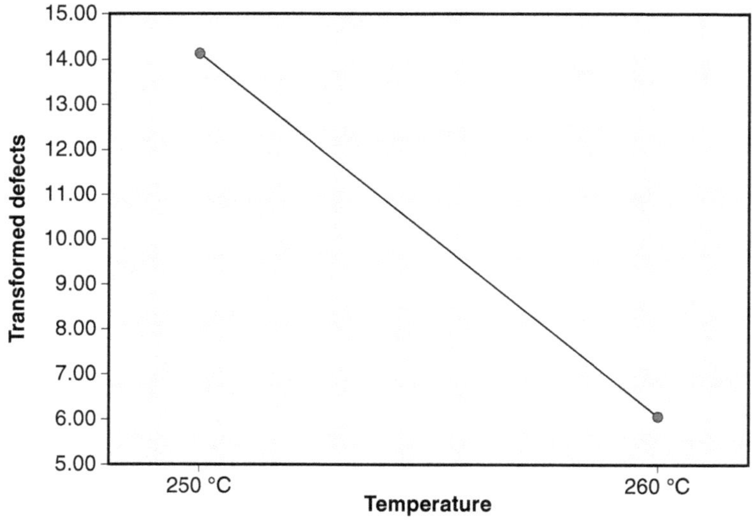

Figure 6.9 Plot of effect B (temperature) on transformed defects.

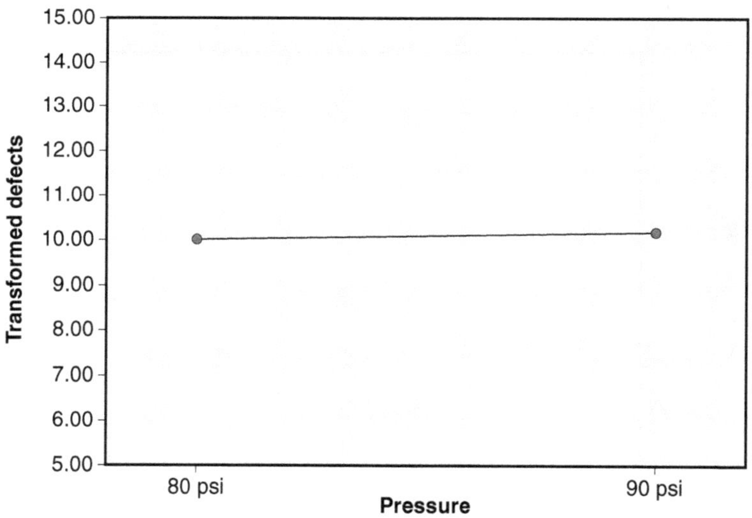

Figure 6.10 Plot of effect C (pressure) on transformed defects.

$$C_{+,-} = \bar{Y}_{+,-} - \bar{\bar{Y}}$$

where

$\bar{\bar{Y}}$ = Average run S/N ratio

$\bar{Y}$ = Average at a treatment level

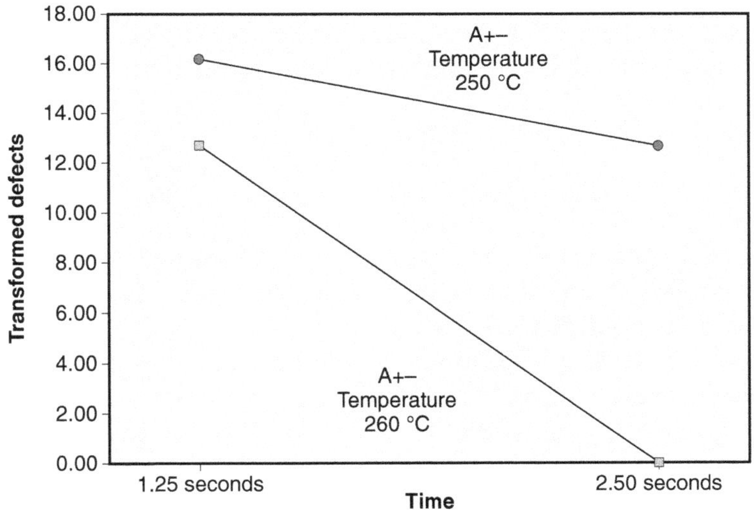

Figure 6.11 Plot of interaction AB (time-temperature) on transformed defects.

$A_+ = 5.62 - 10.06 = -4.44$

$A_- = 14.50 - 10.06 = 4.44$

$B_+ = 6.06 - 10.06 = -4.00$

$B_- = 14.06 - 10.06 = 4.00$

$C_+ = 10.13 - 10.06 = 0.06$

$C_- = 9.99 - 10.06 = -0.07$

$AB_+ = 8.43 - 10.06 = -1.63$

$AB_- = 11.69 - 10.06 = 1.63$

$AC_+ = 9.40 - 10.06 = -0.66$

$AC_- = 10.72 - 10.06 = 0.66$

$BC_+ = 10.42 - 10.06 = 0.36$

$BC_- = 9.70 - 10.06 = -0.36$

$ABC_+ = 10.43 - 10.06 = 0.37$

$ABC_- = 9.69 - 10.06 = -0.37$

Since our objective is to minimize defects, set factor A (time) to 2.50 seconds and factor B (temperature) to 260 °C. Refer to Figure 6.11.

$\hat{Y} = \overline{\overline{Y}} +$ Coefficients of the significant effects, significant interactions, and individual effects from significant interactions

$$\hat{Y} = \overline{\overline{Y}} + A_{+} + B_{+} + AB_{+}$$

$$\hat{Y} = -4.44 + -4.00 + -1.56 = -10.00$$

To transpose to original units, take the sine of –10 and square it. The prediction in the original units is .03% or three defects. Please note that Figure 6.11 indicates there should be zero defects; the difference is due to rounding.

At this point it would be wise to verify experimental results by using the model of the experiment to operate the process. All nonsignificant factors can be set solely on the basis of cost, productivity, or convenience. One may also consider opening up the window of the significant factors to see if further process improvement is possible. Remember that most problems are solved by a sequence of studies that move toward an optimum region. This requires that one be aware of physical and safety constraints that may limit further studies. If still not satisfied with the performance, look for other factors.

6.3 ORDERED CATEGORICAL DATA EXAMPLE

Another method for using qualitative measurement is *ordered categorical data*. This method utilizes a *weighted probability scoring scheme* (WPPS) applied to ordered categorical data. Again, the importance of a good measurement systems analysis (MSA), test method validation (TMV), or gage repeatability and reproducibility (GRR) study cannot be understated for ordered categorical output responses, and must be considered a prerequisite to executing a DOE.

A manufacturing company produces a part that has a quality characteristic that is assessed with ordered categorical data using a weighted probability scoring scheme as shown in Table 6.3.

A production process has been generating product with varying levels of quality. The defect cannot be measured using quantitative (variable) data and is assessed with ordered categorical data using a weighted probability scoring scheme. The goal of the

Table 6.3 Quality characteristic scoring scheme.

Rating	Criteria	Weight
I	Excellent	10
II	Very good	8
III	Good	6
IV	Fair	4
V	Poor	2

experiment is to improve the product quality. The process has three key variables: time, temperature, and pressure. The team decided, based on experience and the manufacturer's recommendation, to use the following levels:

A: Time—1.25 seconds and 2.50 seconds

B: Temperature—250 °C and 260 °C

C: Pressure—80 psi and 90 psi

Ten appraisers evaluated the part produced for each run using the criteria in Table 6.3. The results are shown in Table 6.4.

Note: It should be obvious from this example that the measurement systems analysis (MSA), test method validation (TMV), or gage repeatability and reproducibility (GRR) study was not adequate for the evaluation of the feature. The results shown are to demonstrate the procedure for using ordered categorical data using WPPS.

Convert each ranking to a probability by simply dividing the number of observations for each ranking by the number of rankings. For run 1, category 1:

$$p = \frac{5}{10} = 0.50$$

Note: Five of the 10 appraisers rated run 1 as category 1 (5).

See Table 6.5 for the completed conversions.

Calculate W for each run using the following formula:

$$W(\text{Run}) = w_1\text{pI} + w_2\text{pII} + w_3\text{pIII} + w_4\text{pIV} + w_5\text{pV}$$

where

w = Weight factor

p = Probability.

$$W(\text{Run1}) = 10(0.05) + 8(0.30) + 6(0.20) + 4(0.00) + 2(0.00) = 8.60$$

$$W(\text{Run2}) = 10(0.30) + 8(0.20) + 6(0.20) + 4(0.20) + 2(0.10) = 6.80$$

$$W(\text{Run3}) = 10(0.10) + 8(0.30) + 6(0.40) + 4(0.10) + 2(0.10) = 6.40$$

$$W(\text{Run4}) = 10(0.10) + 8(0.10) + 6(0.30) + 4(0.30) + 2(0.20) = 5.20$$

$$W(\text{Run5}) = 10(0.40) + 8(0.30) + 6(0.20) + 4(0.10) + 2(0.00) = 8.00$$

$$W(\text{Run6}) = 10(0.10) + 8(0.20) + 6(0.40) + 4(0.20) + 2(0.10) = 6.00$$

$$W(\text{Run7}) = 10(0.30) + 8(0.20) + 6(0.20) + 4(0.20) + 2(0.10) = 6.80$$

$$W(\text{Run8}) = 10(0.20) + 8(0.10) + 6(0.20) + 4(0.30) + 2(0.02) = 5.60$$

Table 6.4 Ordered categorical data example.

Run	A Time	B Temp	C Pressure	AB Interaction	AC Interaction	BC Interaction	ABC Interaction	I	II	III	IV	V
1	–	–	–	+	+	+	–	5	3	2	0	0
2	+	–	–	–	–	+	+	3	2	2	2	1
3	–	+	–	–	+	–	+	1	3	4	1	1
4	+	+	–	+	–	–	–	1	1	3	3	2
5	–	–	+	+	–	–	+	4	3	2	1	0
6	+	–	+	–	+	–	–	1	2	4	2	1
7	–	+	+	–	–	+	–	3	2	2	2	1
8	+	+	+	+	+	+	+	2	1	2	3	2

Table 6.5 Observations converted to probabilities.

	Observations					Probabilities					
Run	**I**	**II**	**III**	**IV**	**V**	**I**	**II**	**III**	**IV**	**V**	**W**
1	5	3	2	0	0	0.50	0.30	0.20	0.00	0.00	8.60
2	3	2	2	2	1	0.30	0.20	0.20	0.20	0.10	6.80
3	1	3	4	1	1	0.10	0.30	0.40	0.10	0.10	6.40
4	1	1	3	3	2	0.10	0.10	0.30	0.30	0.20	5.20
5	4	3	2	1	0	0.40	0.30	0.20	0.10	0.00	8.00
6	1	2	4	2	1	0.10	0.20	0.40	0.20	0.10	6.00
7	3	2	2	2	1	0.30	0.20	0.20	0.20	0.10	6.80
8	2	1	2	3	2	0.20	0.10	0.20	0.30	0.20	5.60

Nine Steps for Analysis of Effects

1. *Calculate the absolute values of the effects*:

$$|E(A)| = |\overline{Y}_{A+}| - |\overline{Y}_{A-}|$$

$$= \frac{12.70+0.00+9.79+0.00}{4} - \frac{16.26+11.54+17.47+12.71}{4} = 1.55$$

$$|E(B)| = |\overline{Y}_{B+}| - |\overline{Y}_{B-}|$$

$$= \frac{11.54+0.00+12.71+0.00}{4} - \frac{16.26+12.70+17.47+9.79}{4} = 1.35$$

$$|E(C)| = |\overline{Y}_{C+}| - |\overline{Y}_{C-}|$$

$$= \frac{16.26+12.70+11.54+0.00}{4} - \frac{17.47+9.79+12.71+0.00}{4} = 0.15$$

$$|E(AB)| = |\overline{Y}_{AB+}| - |\overline{Y}_{AB-}|$$

$$= \frac{16.26+0.00+17.47+0.00}{4} - \frac{12.70+11.54+9.79+12.71}{4} = 0.35$$

$$|E(AC)| = |\overline{Y}_{AC+}| - |\overline{Y}_{AC-}|$$

$$= \frac{16.26+11.54+9.79+0.00}{4} - \frac{12.70+0.00+17.47+12.71}{4} = 0.05$$

$$|E(BC)| = |\overline{Y}_{BC+}| - |\overline{Y}_{BC-}|$$

$$= \frac{16.26+12.70+12.71+0.00}{4} - \frac{11.54+0.00+17.47+9.79}{4} = 0.55$$

$$|E(ABC)| = |\overline{Y}_{ABC+}| - |\overline{Y}_{ABC-}|$$

$$= \frac{12.70+11.54+17.47+0.00}{4} - \frac{16.26+0.00+9.79+12.71}{4} = 0.05$$

Note: The minus signs have been removed for simplicity.

2. *Create a Pareto chart of the absolute values of the effects.* The Pareto chart in Figure 6.12 indicates that factors A (time) and B (temperature) may be significant. The interaction of BC (temperature and pressure) may also be significant.

3. *Calculate the standard deviation of the experiment.* This step does not apply with qualitative data.

Table 6.6 Table of effects.

Run	A Time	B Temp	C Pressure	AB Interaction	AC Interaction	BC Interaction	ABC Interaction	W
1	–	–	–	+	+	+	–	8.60
2	+	–	–	–	–	+	+	6.80
3	–	+	–	–	+	–	+	6.40
4	+	+	–	+	–	–	–	5.20
5	–	–	+	+	–	–	+	8.00
6	+	–	+	–	+	–	–	6.00
7	–	+	+	–	–	+	–	6.80
8	+	+	+	+	+	+	+	5.60
$\Sigma\bar{Y}+$	23.60	24.00	26.40	27.40	26.60	27.80	26.80	
$\Sigma\bar{Y}-$	29.80	29.40	27.00	26.00	26.80	25.60	26.60	
$\bar{Y}+$	5.90	6.00	6.60	6.85	6.65	6.95	6.70	$\bar{\bar{Y}}$ 6.68
$\bar{Y}-$	7.45	7.35	6.75	6.50	6.70	6.40	6.65	
C+	–0.78	–0.68	–0.08	0.17	–0.03	0.27	0.02	
C–	0.77	0.67	0.07	–0.18	0.02	–0.28	–0.03	
Abs effect	1.55	1.35	0.15	0.35	0.05	0.55	0.05	

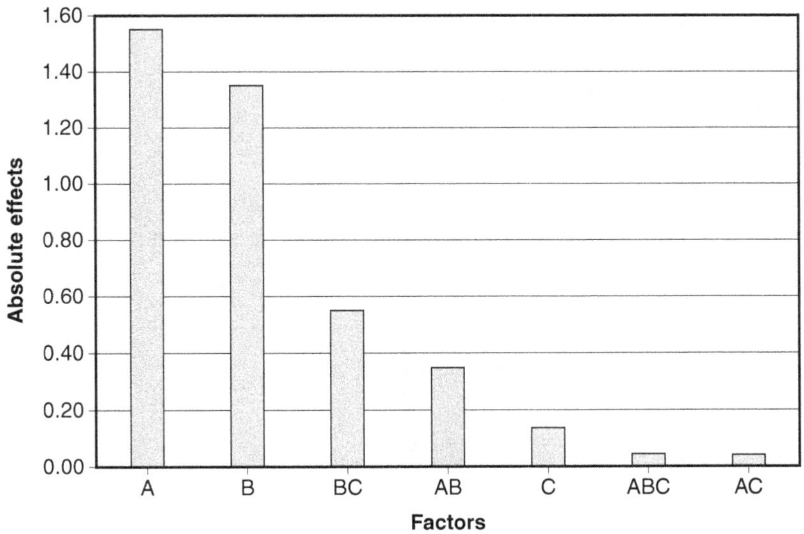

Figure 6.12 Pareto chart of the absolute values of the effects.

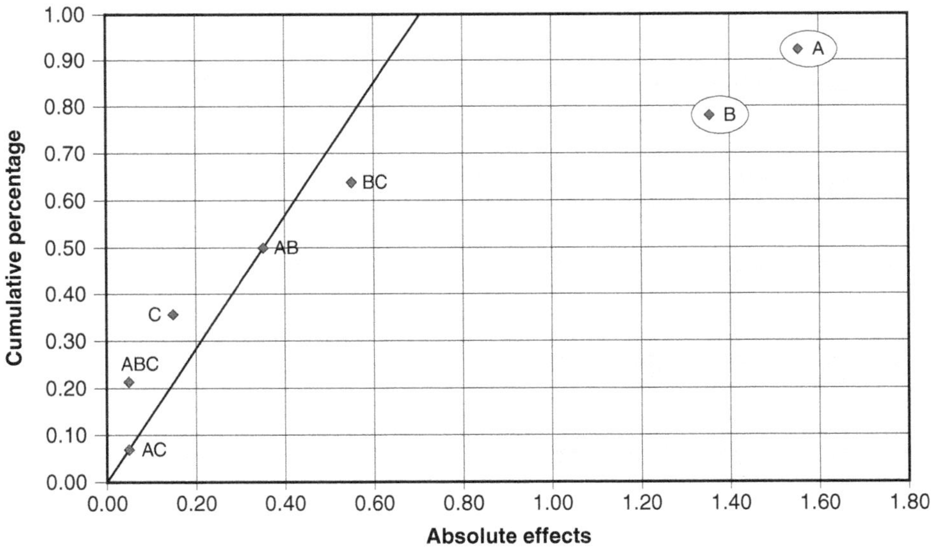

Figure 6.13 Half-normal plot of the absolute effects.

4. *Calculate the standard deviation of the effects.* This step does not apply with qualitative data.

5. *Determine the statistic.* This step does not apply with qualitative data.

6. *Calculate the decision limits.* This step will be done with the use of a half-normal plot (Figure 6.13).

7. *Determine the significant effects.* An analysis of Figure 6.13 indicates that factors A (time) and B (temperature) are significant.

8. *Graph the main effects and any significant interactions* (Figures 6.14, 6.15, and 6.16).

9. *Model the significant effects, significant interactions, and individual effects from significant interactions.* Since our objective is to improve product quality, factor A (time

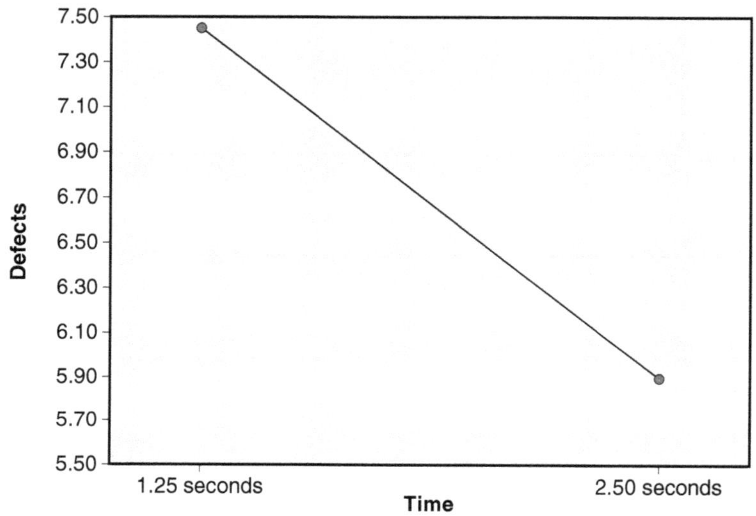

Figure 6.14 Plot of effect A (time).

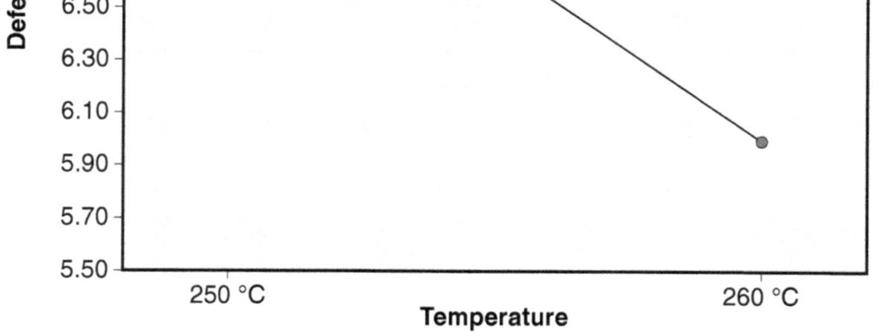

Figure 6.15 Plot of effect B (temperature).

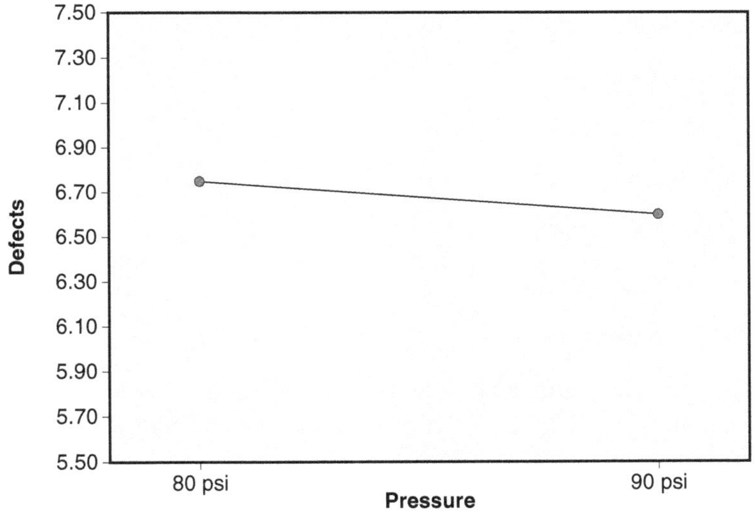

Figure 6.16 Plot of effect C (pressure).

1.25 seconds or 2.50 seconds) and factor B (temperature 250 °C or 260 °C) are set at the low (–1) levels. These levels will provide the lowest number of defects. Factor C (pressure 80 psi or 90 psi) is not significant and can be set solely on the basis of cost, productivity, or convenience.

The coefficients must be calculated so the model can be compiled. The coefficient calculations are simply the average difference of the overall average from a level (+, –). Note that it is not necessary to calculate all of the coefficients. All coefficients are calculated for illustrative purposes.

$$C_{+,-} = \bar{Y}_{+,-} - \bar{\bar{Y}}$$

where

$\bar{\bar{Y}}$ = Average run S/N ratio

$\bar{Y}$ = Average at a treatment level

$A_+ = 5.90 - 6.68 = -0.78$

$A_- = 7.45 - 6.68 = 0.77$

$B_+ = 6.00 - 6.68 = -0.68$

$B_- = 7.35 - 6.68 = 0.67$

$C_+ = 6.60 - 6.68 = -0.08$

$C_- = 6.75 - 6.68 = 0.08$

$AB_+ = 6.85 - 6.68 = 0.17$

$$AB_{-} = 6.50 - 6.68 = -0.18$$

$$AC_{+} = 6.65 - 6.68 = -.03$$

$$AC_{-} = 6.70 - 6.68 = 0.02$$

$$BC_{+} = 6.95 - 6.68 = 0.27$$

$$BC_{-} = 6.40 - 6.68 = -0.28$$

$$ABC_{+} = 6.70 - 6.68 = 0.02$$

$$ABC_{-} = 6.65 - 6.68 = -0.03$$

Since our objective is to improve product quality, set factor A (time) to 1.25 seconds and factor B (temperature) to 250 °C. Refer to Figures 6.14 and 6.15. Factor C (pressure 80 psi or 90 psi) is not significant and can be set solely on the basis of cost, productivity, or convenience.

$\hat{Y} = \bar{\bar{Y}} +$ Coefficients of the significant effects, significant interactions, and individual effects from significant interactions

$$\hat{Y} = \bar{\bar{Y}} + A_{+} + B_{+}$$

$$\hat{Y} = 6.68 + 0.77 + 0.67 = 8.12$$

The predicted weighted response is 8.12 when factor A (time) is 1.25 seconds and factor B (temperature) is 250 °C. Factor C (pressure 80 psi or 90 psi) is not significant and can be set solely on the basis of cost, productivity, or convenience.

At this point it would be wise to verify experimental results by using the model of the experiment to operate the process. All nonsignificant factors can be set solely on the basis of cost, productivity, or convenience. One may also consider opening up the window of the significant factors to see if further process improvement is possible. Remember that most problems are solved by a sequence of studies that move toward an optimum region. This requires that one be aware of physical and safety constraints that may limit further studies. If still not satisfied with the performance, look for other factors.

7

Screening and Other Designs

The necessary analytical techniques have now been covered. They will work for any number of factors in a DOE as long as only two levels are used. As the number of factors increases in a factorial design, the cost and difficulty of control increase even faster due to the acceleration in the number of trials required. It's necessary now to study the design techniques that economically permit DOEs with many factors. These are the *screening designs* that can be used for as few as four factors or as many as can be practically handled or afforded in an experiment. Their purpose is to identify or screen out those factors that merit further investigation. To provide some idea of the power that is being introduced, consider Table 7.1.

Can you imagine the budget for a 2^{31} factorial experiment? Now, that is job security! If you performed one experiment per minute, 24 hours a day, 365 days per year, it would take 4,085.8 years to complete the DOE! Even a much smaller 2^7 factorial requires 128 trials—still an expensive investment for most processes. The comparative numbers required by screening designs are not typographical errors; they represent the smallest experiment that can provide substantial information. Looks too good to be true? You are

Table 7.1 Comparison of the number of runs in factorial and screening designs.

Number of factors	Full factorial	Minimum screening design
4	16	8
5	32	8
6	64	8
7	128	8
11	2,048	12
15	32,768	16
19	524,288	20
23	8,388,868	24
27	134,217,728	28
31	2,147,483,648	32

right—there are penalties to be paid for this economy. But the penalties are minor compared to the immense advantages to be gained.

7.1 CONFOUNDING, ALIASES, AND RESOLUTION

Before going further, the concepts of *confounding* and *aliases* must be defined. Consider the analysis matrix for a factorial experiment with three factors (Table 7.2).

Assume that there actually are four factors to be studied, requiring 16 runs for a full factorial. Assume further that, due to material limitations or cost, only eight runs can be performed. How could an extra factor be packed into this design? Why not put factor D in the ABC column and treat that column like a factor? It has already been said that a three-factor interaction is not realistic, so it probably wouldn't be missed. What are the penalties of this move? The penalty is that each two-factor interaction is now identical with another two-factor interaction. For example, calculate a column for CD and compare it to the AB column. They are identical, and AB = CD. This is illustrated in Table 7.3.

Table 7.2 Analysis table for three factors.

Run	A	B	C	AB	AC	BC	ABC = D
1	–	–	–	+	+	+	–
2	+	–	–	–	–	+	+
3	–	+	–	–	+	–	+
4	+	+	–	+	–	–	–
5	–	–	+	+	–	–	+
6	+	–	+	–	+	–	–
7	–	+	+	–	–	+	–
8	+	+	+	+	+	+	+

Table 7.3 Illustration of identical interactions.

Run	A	B	AB	C	D	CD
1	–	–	+	–	–	+
2	+	–	–	–	+	–
3	–	+	–	–	+	–
4	+	+	+	–	–	+
5	–	–	+	+	+	+
6	+	–	–	+	–	–
7	–	+	–	+	–	–
8	+	+	+	+	+	+

When the AB effect is calculated, the experimenter does not know whether that value is the AB effect, the CD effect, or a combination of both. This confusing of effects is called *confounding*, and it is said that AB is *confounded* with CD. The confounded effects are also called *aliases*; aliases are identical except for sign. If AB and CD are aliases, their sign column in an analysis matrix can be either identical or opposite in sign. The same is true of their calculated effects. The decision to treat factor D as if it were the ABC column generates an identity that can be used to define the aliases. That is, the aliases are defined as D = ABC. The rule for manipulating such relationships follows.

For any defined relationship, a squared term will equal unity or 1; for example, if both sides of the previous relationship are multiplied by D, the result is

$$D \times D = D^2 = 1 = ABCD.$$

This is the defining relationship or identity that defines the aliases when a fraction of the full factorial design is performed. Generally, the 1 is replaced with an "I." To determine the aliases for any effect, multiply both sides of the identity by that effect. For this example, the identity is

$$I = ABCD.$$

Multiplying both sides by AB will provide the alias for AB:

$$IAB = AB = A^2B^2CD = CD$$

Thus, AB is confounded with CD; one can also state that AB and CD are aliases.

Table 7.4 shows all the aliases for this design, which is a half-fractional factorial (half the number of runs for a full factorial).

As more factors are packed into a design, the definition of the identities becomes more difficult. Therefore, the aliases are provided or described in all of the screening designs used by this author. Confounding is the penalty paid for using fewer runs than a full factorial in a design. While there are times when this penalty is very troublesome, most of the time it is a risk that can be managed, and rules are covered to reduce this risk further. In any screening design, one must be aware of the confounding pattern.

Resolution refers to the ability to separate main effects and low-order interactions from one another. The most significant fractional designs are those of resolution III, IV, and V. Resolutions below III are not useful, and resolutions above V are wasteful in that the expanded experimentation has no practical benefit in most cases—the bulk of

Table 7.4 Aliases for a half-fractional factorial design with four factors.

A = BCD	AB = CD
B = ACD	AC = BD
C = ABD	AD = BC

the additional effort goes into the estimation of very high-order interactions that rarely occur in practice. The resolution described in Table 7.5 is only used for regular designs. Regular designs have run size that equals a power of 2, and only full aliasing is present. Non-regular designs are those where run size is a multiple of 4; these designs introduce partial aliasing.

Next, the intent of a screening design must be reviewed. Such designs are for sifting through a large number of factors and selecting those that merit more detailed study. In screening designs, the primary interest is main effects. The levels are set broadly to increase the chance of detecting significant factors. If interactions exist, it is great if they can be identified, but the most important consideration is that they not mislead identification of the factors for further study. The reason for this concern is that in the larger screening designs (five factors and larger), main effects will be confounded with two-factor interactions, either completely or partially. A technique for separating main effects from this confounding with two-factor interactions is discussed later, but even then, two-factor interactions remain confounded with one another.

Table 7.5 Experiment resolution.

Resolution	Ability	Example
I	Not useful: an experiment of exactly one run only tests one level of a factor and hence cannot distinguish between the high and low levels of that factor.	2^{1-1} with defining relation I = A
II	Not useful: main effects are confounded with other main effects.	2^{2-1} with defining relation I = AB
III	Estimate main effects, but these may be confounded with two-factor interactions.	2^{3-1} with defining relation I = ABC
IV	Estimate main effects not confounded by two-factor interactions. Estimate two-factor interaction effects, but these may be confounded with other two-factor interactions.	2^{4-1} with defining relation I = ABCD
V	Estimate main effects not confounded by three-factor (or less) interactions. Estimate two-factor interaction effects not confounded by two-factor interactions. Estimate three-factor interaction effects, but these may be confounded with other two-factor interactions.	2^{5-1} with defining relation I = ABCDE
VI	Estimate main effects not confounded by four-factor (or less) interactions. Estimate two-factor interaction effects not confounded by three-factor (or less) interactions. Estimate three-factor interaction effects, but these may be confounded with other three-factor interactions.	2^{6-1} with defining relation I = ABCDEF
VII	Estimates of main factors, two-factor and three-factor effects are not confounded with one another but may be confounded with higher-order interactions. Four-factor and higher interactions may be confounded.	2^{7-1} with defining relation I = ABCDEFG

Source: Adapted from http://en.wikipedia.org/wiki/Fractional_factorial_design.

Recall that the order of studying designs was said to be backward. For ease of instruction, the author started with small numbers of factors and progressed to large numbers of factors. In practice, a study begins with a screening design. It is the tool used to sift through many potential factors, identifying those that need to be studied in smaller refining experiments in which interactions can be more thoroughly assessed.

7.2 SCREENING DESIGNS

Screening designs can be used to study $n - 1$ factors in n runs in which n is divisible by 4. That is, seven factors can be studied in eight runs, 11 factors can be studied in 12 runs, 15 factors in 16 runs, and so on. Such designs are of two distinctly different types:

1. *Geometric designs.* These designs are those in which n is a power of 2. That is, the number of runs is eight, 16, 32, 64, and so on. These designs may be fractional factorials or Plackett-Burman designs. They are the same if a maximum number of factors is used in the design. In the geometric designs, the confounding that exists is complete. In other words, if two effects are confounded, they are identical except possibly for sign. (If AB and CD are confounded, their effects are numerically identical but one may be the negative of the other.) It is easy to specify precisely what is confounded with what (the aliases), and it should be done. This perfect confounding is the advantage of geometric designs in that one can identify the risks in the interpretation of effects.
2. *Nongeometric designs.* These are designs that are divisible by 4 but are not powers of 2. Such designs have runs of 12, 20, 24, 28, 36, and so on. These designs are Plackett-Burman designs and represent a unique contribution to the available designs. These designs do not have complete confounding of effects. Instead, each main effect is confounded partially with all interactions that do not contain that main effect. This is both an advantage and a disadvantage. The advantage is that all of this partial confounding tends to average out to a negligible impact, and it may be possible to get some information on a significant interaction. The disadvantage is that the presence of a large interaction may distort the estimated effects of several individual factors since each interaction is partially confounded with all main effects except the two interacting factors.

This discussion touches the tip of a technical iceberg that is deferred to the study of more-advanced references. One should note that while fractional factorial designs are more difficult to portray, software makes them very easy to use. Fractional factorials assure efficient control of the confounding pattern, whereas Plackett-Burman designs are easiest to apply. This comparison becomes a concern only when a design is performed with less than the maximum number of possible factors. Plackett-Burman designs are provided for six or more factors. Fractional factorial designs are provided for four- and five-factor experiments in which a full factorial is not practical. (Special designs are provided for these two cases since there is greater efficiency in using fractional factorials for them than in using Plackett-Burman designs.) Designs are listed in Appendix J. There are two critical characteristics of screening designs that help make the screening designs work despite confounding:

1. *Effect heredity.* Large main effects often have significant interactions. This provides guidelines on where to look for interactions.
2. *Effect sparsity.* Main effects are generally much larger than two-factor interactions, while three-factor and higher-order interactions can generally be ignored. This means that main effects generally dominate interactions, and it allows the experimenter to ignore three-factor (and higher) interactions as rare events.

The eight-step analysis procedure for effects and the variation procedure are applied in screening designs just as they have been applied in previous examples.

7.3 REFLECTION

A *reflection* (called a *fold-over* design in some references) is the exact opposite of the basic design in signs. It is achieved by replacing all plus signs with minus signs, and vice versa. The reflected design reverses the signs of the two-factor interactions in the confounding pattern. Let me describe the meaning of the confounding pattern more precisely using only factor A of the current design. When it is stated that the confounding is such that A = –BD = –CG = –EF, this really means that the estimated effect of A is a combination of the actual values of the effects of A, BD, CG, and EF in the relationship shown in the base design:

$$E(A)_{Est} = E(A)_{Act} - E(BD)_{Act} - E(CG)_{Act} - E(EF)_{Act}.$$

The reflection provides a confounding pattern opposite in sign such that the reflected design is

$$E(A)_{Est} = E(A)_{Act} + E(BD)_{Act} + E(CG)_{Act} + E(EF)_{Act}.$$

If the results of a reflected design are added to those of a base design and the average calculated, the two equations show that the interaction terms cancel, leaving an estimate of main effects free from confounded two-factor interactions:

$$\tfrac{1}{2}\,\{\text{Base } E(A)_{Est} + \text{Reflected } E(A)_{Est}\} = \tfrac{1}{2}\,\{[E(A)_{Est} - E(BD)_{Act} - E(CG)_{Act} - E(EF)_{Act}] + [E(A)_{Act} + E(BD)_{Act} + E(CG)_{Act} + E(EF)_{Act}]\}$$
$$= \tfrac{1}{2}\{E(A)_{Act} + E(A)_{Act}\} = E(A)_{Act}$$

This is the characteristic and purpose of reflection in screening designs: combining a reflection with a base design will always provide estimates of main effects that are free of two-factor interactions. The two-factor interactions will still be confounded with one another. In the current example, to determine whether the three main effects are realistic or whether some main effect–interaction combination is the actual model, one needs only to run a reflection.

Before proceeding, there is one other aspect of the foregoing relationships that can be used to provide information. If averaging a reflection with the base design clarifies main effects, what if one took the average differences of the effects, that is, {reflected $E(A)_{Est}$ – Base $E(A)_{Est}$ }/2? These relationships show that such a calculation estimates the combined effects of the confounded interactions:

$$\tfrac{1}{2}\,[\text{Reflected } E(A)_{Est} - \text{Base } E(A)_{Est}] = \tfrac{1}{2}\{[E(A)_{Act} + E(BD)_{Act} + E(CG)_{Act} + E(EF)_{Act}] - [E(A)_{Act} - E(BD)_{Act} - E(CG)_{Act} - E(EF)_{Act}]\} = \tfrac{1}{2}\{2\,E(BD)_{Act} + 2\,E(CG)_{Act} + 2\,E(EF)_{Act}\} = E(BD)_{Act} + E(CG)_{Act} + E(EF)_{Act}$$

While the individual interactions cannot be identified, this calculation at least gives some indication of the existence of interactions.

This is more easily demonstrated than described by analyzing a reflection of the current design and reanalyzing the results. As stated earlier, the average of the two sets of effects (base and reflected) provides estimates of main effects that are free of any two-factor interaction effects (see Table 7.6).

The averages of the combined estimates provide clean estimates of main effects, and the conclusion is that the effects of C and E remain significant and consistent, while the effect of B cannot be distinguished from the inherent variation. Factor B apparently indicates a confounded interaction. The average difference row confirms this, as already discussed, by revealing that the group of interactions confounded with B is significant. Probably only one of the interactions is significant. This group is AD, CE, and FG. The heredity rule from the two critical characteristics of screening designs states that significant main effects often have significant interactions. The most likely candidate is CE since both C and E stand confirmed as significant. This should be verified in a subsequent refining experiment in which the interaction can be assessed directly.

Table 7.6 Summary of effects.

Factors	**A**	**B**	**C**	**D**	**E**	**F**	**G**
Base design (b)	1.13	10.39	24.89	–1.32	14.78	0.29	–0.06
Reflection (r)	0.89	–2.69	18.87	1.29	15.81	1.37	0.65
Average = (b + r) / 2	1.01	3.85	21.88	–0.02	15.30	0.83	0.30
Difference = (r – b) / 2	–0.12	–6.54	–3.01	1.31	0.52	0.54	0.36

7.4 OTHER ANALYTICAL CONSIDERATIONS

Replications are expensive. By the time this eight-run Plackett-Burman design was reflected and replicated, 32 trials were required. What if one could not afford to replicate? Could the DOE still be analyzed? Yes, simply working from the Pareto analysis would have been very effective. Tests of significance are nice but by no means essential. In general, if one has a choice of reflection or replication, reflection should be used. The exception to this rule arises when there is reason to believe that the experiment will be subject to special cause variation or outliers. For a variation analysis, the technique introduced earlier using residuals can be applied when there is no replication.

In review, the basic screening designs are highly efficient for sifting through many factors to select a smaller subset for more detailed study. Their penalty is the confounding of main effects with two-factor interactions—or at least the confounding of interactions with one another. By doubling the number of trials in the experiment, reflection assures clean assessment of main effects with limited information on interactions. Screening designs are a logical launch point for a sequence of experiments that may culminate in optimization studies of a very few factors. In addition, note how easily one can identify an even larger group of factors that are not worth further study. This allows the experimenters to address situations where money is being spent to control factors that do not matter.

7.5 EVEN LARGER DESIGNS

These principles of design and analysis are used with even larger designs. The 16-run, in particular, is a major workhorse. The size of the experiment is limited by the ability of the experimenter to afford and control very large experiments. From a mathematical viewpoint, larger is better since there are more terms, which tends to average the impact of confounding to a net zero. From the standpoint of the experimenter who is trying to assure that the correct treatment combinations are performed, there is a point at which control becomes difficult. Cost and control are the realistic limits to the size of an experiment. Appendix I contains designs up to a 20-run Plackett-Burman and provides analysis tables for each. Software will provide even larger designs and provide a choice of Plackett-Burman or fractional factorial designs.

7.6 OTHER TYPES OF DESIGNS

This has been an introduction to the basic designs—those with two levels. Three-level designs can also be generated as either full factorials or as fractional factorials when nonlinearity is a major concern. The analysis of such designs is not as easy as with the two-level designs, and more runs are necessary. There are also some unusual Taguchi designs that combine two-level and three-level factors. One of the more useful of these is the L-18 design, which will analyze one factor at two levels and seven factors at three levels, requiring only 18 runs. Additionally, there are highly efficient designs called *response surface* designs that expand small two-level designs to efficiently address curvature. These are often used to model processes in the final refining or optimization stage. (The subject of optimization designs definitely merits further study.) *Evolutionary operation* (EVOP) is a technique for optimizing a process while it is operating. A very different family of designs is the family of *mixture* designs, which are appropriate when the factors' levels must total to some fixed percentage of the total volume of a material. These and other designs are discussed in advanced DOE training and must be conducted with software to facilitate their analysis.

At this point, the reader must again be reminded that this has been an introduction to DOE. The intent has been to introduce the basic concepts and to encourage you to go on to study more-advanced references. Some significant points have intentionally been skipped or glossed over in the interest of simplicity. The reader is strongly encouraged to review the references provided in the Bibliography to gain exposure to more-advanced techniques and concepts.

8

Taguchi Experiments

Dr. Genichi Taguchi built upon the work of Sir Ronald A. Fischer and others in developing a series of special orthogonal arrays. His modifications to the existing arrays allowed experimenters the ability to conduct the first run in an experiment using the current operating settings. In addition to his development and modifications of orthogonal arrays, Taguchi also introduced the concept of the *loss function* and the use of *signal-to-noise* (S/N) ratios (which are derived from the loss function). Although the examples presented in this chapter will use the S/N ratio to identify the best option to reduce variation, averages and standard deviation analysis techniques presented previously can be used.

Taguchi focused on the three stages of product development to reduce or otherwise eliminate variation: system design, parameter design, and tolerance design. Collectively, these concepts are known as *robust design*. This concept effectively requires the experimenter to apply resources to control the *critical to quality* (CTQ) features or characteristics.

Taguchi taught a two-step optimization process:

1. Reduce the variation in the process (precision)
2. Adjust the process to the target value (accuracy)

Figure 8.1 graphically depicts the difference between accuracy and precision. *Accuracy* defines how close the values are to the target value. *Precision* is a measure of the dispersion. Figure 8.2 depicts the adjustment of the process to the target value.

Taguchi's view is that all factors are either control factors or noise factors. *Control factors* affect the system, and can be set by the experimenter. *Noise factors* affect the system but are difficult to control, such as environmental factors including temperature and humidity (see Figure 8.3).

8.1 TAGUCHI ORTHOGONAL DESIGNS

Taguchi orthogonal arrays typically use numbers for the columns instead of the traditional use of letters. Additionally, factor levels are labeled 1, 2, 3, . . . instead of the customary use of + and – signs because some designs allow the use of three or more levels. Taguchi generally considers factor interactions as noise and therefore not significant.

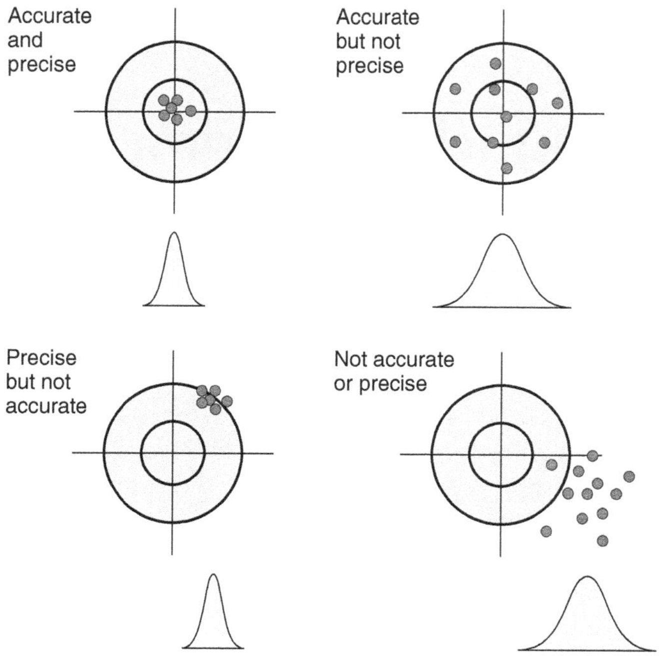

Figure 8.1 Accuracy versus precision.

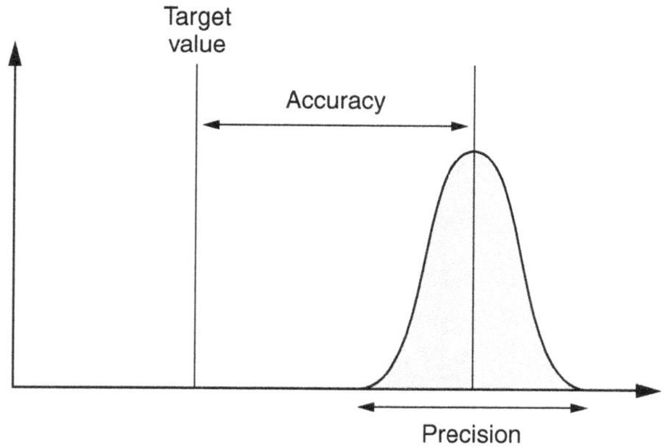

Figure 8.2 Adjusting the process to the target value.

However, the Taguchi orthogonal designs do allow for the study of interactions. Figure 8.4 shows a Taguchi L4 orthogonal array and interaction table.

The interaction table is read by row and column. For example, if the experimenter wishes to study the interaction of factor 1 and factor 2, the interaction column would be 3 (see Figure 8.5).

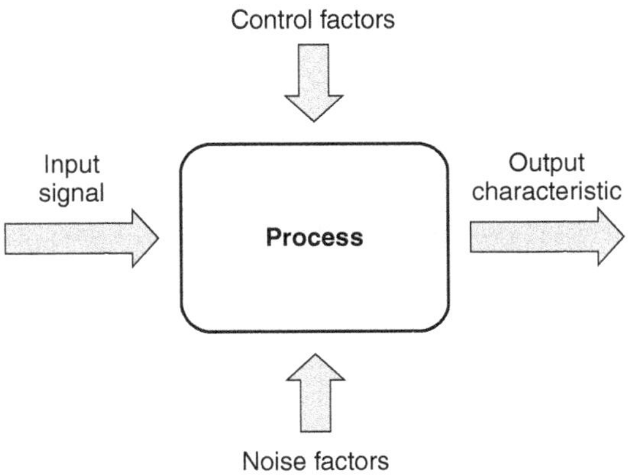

Figure 8.3 Taguchi's view of a process.

	1	**2**	**3**
1	1	1	1
2	1	2	2
3	2	1	2
4	2	2	1

	1	**2**	**3**
1		3	2
2			1

Figure 8.4 Taguchi L4 orthogonal array and interaction table.

If the experimenter wishes to study the interaction of factor 2 and factor 3, the interaction column would be 1 (see Figure 8.6).

The same method is used to determine the interaction column for Taguchi arrays for which interaction tables are available (see Figure 8.7).

When an experiment does not require the use of all columns, the unused columns are simply ignored. For example, an L4 array is to be used for two factors at two levels, therefore, the column that would normally be used for factor 3 is ignored (see Figure 8.8).

	1	**2**	**3**
1		3	2
2			1

Figure 8.5 Determining the L4 interaction between factors 1 and 2.

	1	**2**	**3**
1		3	2
2			1

Figure 8.6 Determining the L4 interaction between factors 2 and 3.

	1	**2**	**(1, 2)**
1	1	1	2
2	1	2	1
3	2	1	1
4	2	2	2

Figure 8.7 L4 array displaying the interaction column of factors 1 and 2.

	1	**2**	**3**
1	1	1	1
2	1	2	2
3	2	1	2
4	2	2	1

Figure 8.8 Taguchi L4 orthogonal array using only two factors that does not require column 3.

8.2 SIGNAL-TO-NOISE (S/N) RATIOS

Signal-to-noise (S/N) combines the effect of the average and standard deviation into one value used to gage the process output. Taguchi developed several S/N ratios, including:

- *Nominal is best* is used for a continuous nonnegative number that has a specified nominal or target value. The goal is to *maximize* the S/N ratio. Please note that the variance or standard deviation cannot be zero.

$$\text{S/N} = 10\text{Log}_{10}\left(\bar{Y}^2 / s^2\right)$$

where

$\bar{Y}$ = Mean

s = Standard deviation

Example: Calculate the S/N ratio for nominal is best for a process having a $\bar{Y}$ of 2.5 and s 0.75.

$$\text{S/N} = 10\text{Log}_{10}\left(\bar{Y}^2 / s^2\right) = 10\text{Log}_{10}\left(2.5^2 / 0.75^2\right) = 10.46$$

- *Larger is better* is used for a continuous nonnegative number that has a larger is better requirement. The goal is to *maximize* the S/N ratio.

$$\text{S/N} = -10\text{Log}_{10}\left[\frac{1}{n}\Sigma\left(\frac{1}{y_1^2} + \frac{1}{y_2^2} + \ldots \frac{1}{y_n^2}\right)\right]$$

where

n = Number of response

y = Response

Example: Calculate the S/N ratio for larger is better for an experiment with three responses (10.5, 11.2, and 10.7).

$$\text{S/N} = -10\text{Log}_{10}\left[\frac{1}{n}\Sigma\left(\frac{1}{y_1^2} + \frac{1}{y_2^2} + \ldots \frac{1}{y_n^2}\right)\right]$$

$$\text{S/N} = -10\text{Log}_{10}\left[\frac{1}{3}\Sigma\left(\frac{1}{10.5^2} + \frac{1}{11.2^2} + \frac{1}{10.7^2}\right)\right]$$

$$\text{S/N} = 20.66$$

- *Smaller is better* is used for a continuous nonnegative number that has a smaller is better requirement. The goal is to *maximize* the S/N ratio.

$$S/N = -10\text{Log}_{10}\left[\frac{1}{n}\Sigma\left(y_1^2 + y_2^2 + \ldots y_n^2\right)\right]$$

where

n = Number of response

y = Response

Example: Calculate the S/N ratio for smaller is better for an experiment with three responses (10.5, 11.2, and 10.7).

$$S/N = -10\text{Log}_{10}\left[\frac{1}{3}\Sigma\left(10.5^2 + 11.2^2 + \ldots 10.7^2\right)\right]$$

$$S/N = -20.67$$

- *Zero is best* (signed value) is used for a continuous number that can be positive or negative but is usually zero. The goal is to *maximize* the S/N ratio. Please note that the variance or standard deviation cannot be zero.

$$S/N = -10\text{Log}_{10}\left(\sigma^2\right)$$

where

σ = Standard deviation

Example: Calculate the S/N ratio for zero is best for a process with σ 0.75.

$$S/N = -10\text{Log}_{10}\left(\sigma^2\right) = -10\text{Log}_{10}\left(0.75^2\right) = 2.50$$

- *Fraction defective* is used for attribute data that have a target fraction defective of zero. The goal is to *maximize* the S/N ratio.

$$S/N = -10\text{Log}_{10}\left[P/(1-P)\right]$$

where

P = Proportion defective

Example: Calculate the S/N ratio for a fraction defective for a process with P 0.03.

$$S/N = -10\text{Log}_{10}\left[P/(1-P)\right] = -10\text{Log}_{10}\left[0.03/(1-0.03)\right] = 15.10$$

- *Ordered categorical* is used for rating systems or categorical values. Each category is treated as fraction defective. If a *higher* percentage is desired, then *minimize* the S/N ratio. If a *lower* percentage is desired, then *maximize* the S/N ratio.

8.3 TAGUCHI L4 ORTHOGONAL ARRAY EXAMPLE

A plastic package sealing process has been generating a significant number of complaints due to the excessive force needed to open the packaging. The goal of the experiment is to identify the factor(s) that reduce the force necessary to open the packaging (see Table 8.1). The process has three key variables: time, temperature, and pressure. The team decided, based on experience and the manufacturer's recommendation, to use the following levels:

1: Time—1.25 seconds and 2.50 seconds

2: Temperature—250 °C and 260 °C

3: Pressure—80 psi and 90 psi

Smaller is better is used for a continuous nonnegative number that has a smaller is better requirement. The goal is to *maximize* the S/N ratio.

$$S/N = -10\text{Log}_{10}\left[\frac{1}{n}\Sigma\left(y_1^2 + y_2^2 + \ldots y_n^2\right)\right]$$

where

n = Number of response

y = Response

Table 8.1 Plastic sealing example data.

Run	1 Time	2 Temp	3 Pressure	Y_1	Y_2	Y_3	S/N
1	1	1	1	2.9	3.0	2.8	–9.25
2	1	2	2	2.4	2.2	2.3	–7.24
3	2	1	2	3.5	3.6	3.7	–11.13
4	2	2	1	2.6	2.5	2.7	–8.30
ΣY_1	–16.49	–20.38	–17.55				
ΣY_2	–19.43	–15.54	–18.37				$\bar{\bar{Y}}$ –8.98
$\bar{Y}_1$	–8.25	–10.19	–8.78				
$\bar{Y}_2$	–9.72	–7.77	–9.19				
C_1	0.73	–1.21	0.20				
C_2	–0.74	1.21	–0.21				
Abs effect	1.47	2.42	0.41				

$$S/N_1 = -10\mathrm{Log}_{10}\left[\frac{1}{n}\Sigma\left(y_1^2+y_2^2+y_3^2\right)\right] = -10\mathrm{Log}_{10}\left[\frac{1}{3}\Sigma\left(2.9^2+3.9^2+2.8^2\right)\right]$$
$$= -9.25$$

$$S/N_2 = -10\mathrm{Log}_{10}\left[\frac{1}{n}\Sigma\left(y_1^2+y_2^2+y_3^2\right)\right] = -10\mathrm{Log}_{10}\left[\frac{1}{3}\Sigma\left(2.4^2+2.2^2+2.3^2\right)\right]$$
$$= -7.24$$

$$S/N_3 = -10\mathrm{Log}_{10}\left[\frac{1}{n}\Sigma\left(y_1^2+y_2^2+y_3^2\right)\right] = -10\mathrm{Log}_{10}\left[\frac{1}{3}\Sigma\left(3.5^2+3.6^2+3.7^2\right)\right]$$
$$= -11.13$$

$$S/N_4 = -10\mathrm{Log}_{10}\left[\frac{1}{n}\Sigma\left(y_1^2+y_2^2+y_3^2\right)\right] = -10\mathrm{Log}_{10}\left[\frac{1}{3}\Sigma\left(2.6^2+2.5^2+2.7^2\right)\right]$$
$$= -8.30$$

Nine Steps for Analysis of Effects

1. *Calculate the absolute values of the effects*:

$$|E(1)| = |\bar{Y}_1| - |\bar{Y}_2| = \frac{9.25+7.24}{2} + \frac{11.13+8.30}{2} = 1.47$$

$$|E(2)| = |\bar{Y}_1| - |\bar{Y}_2| = \frac{9.25+11.13}{2} + \frac{7.24+8.30}{2} = 2.42$$

$$|E(3)| = |\bar{Y}_1| - |\bar{Y}_2| = \frac{9.25+8.30}{2} + \frac{7.24+11.13}{2} = 0.41$$

Note: The minus signs have been removed for simplicity.

2. *Create a Pareto chart of the absolute values of the effects.* The Pareto chart in Figure 8.9 indicates that factor 2 (time) and factor 1 (temperature) may be significant.

3. *Calculate the standard deviation of the experiment, s_e.* This step does not apply when using S/N ratios.

4. *Calculate the standard deviation of the effects, s_{Eff}.* This step does not apply when using S/N ratios.

5. *Determine the* t-*statistic.* This step does not apply when using S/N ratios.

6. *Calculate the decision limits.* This step does not apply when using S/N ratios.

7. *Determine the significant effects.* An analysis of Figure 8.9 indicates that factors 1 (temperature) and 2 (time) may be significant.

A half-normal probability plot may not be appropriate for this example because the number of effects is small (three-factor experiment). The results using this method may be erratic. The Pareto chart and factor plots should be evaluated to ensure an appropriate decision is made.

8. *Graph the main effects and any significant interactions* (Figures 8.10, 8.11, and 8.12).

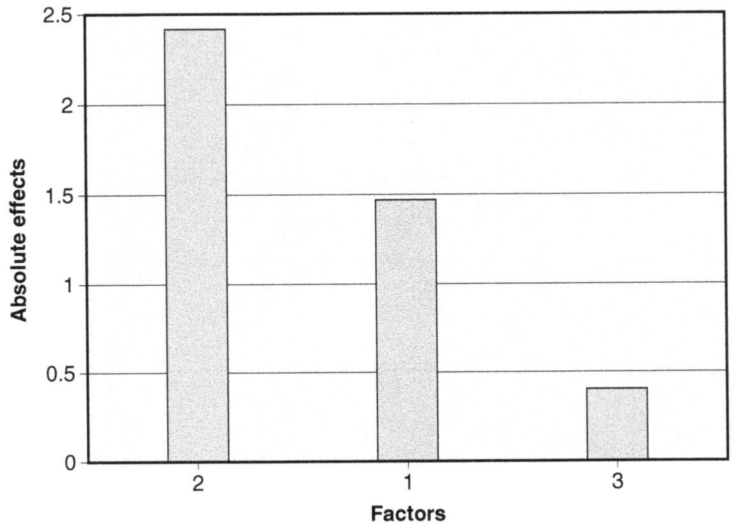

Figure 8.9 Pareto chart of the absolute values of the effects.

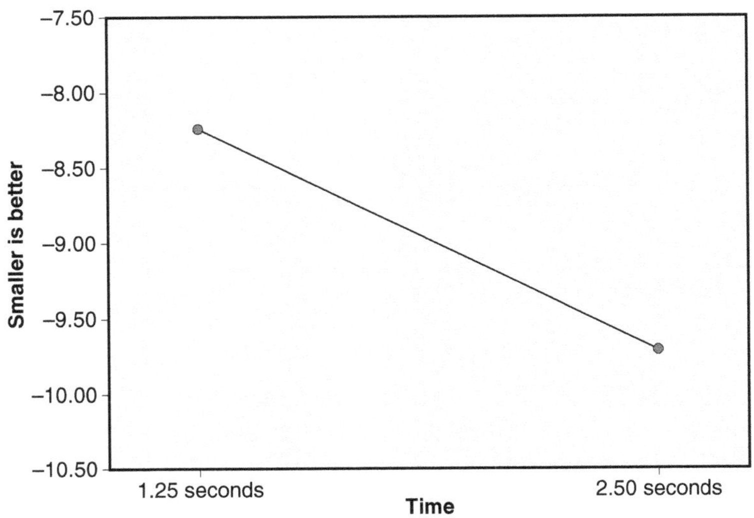

Figure 8.10 Plot of effect 1 (time) on opening force.

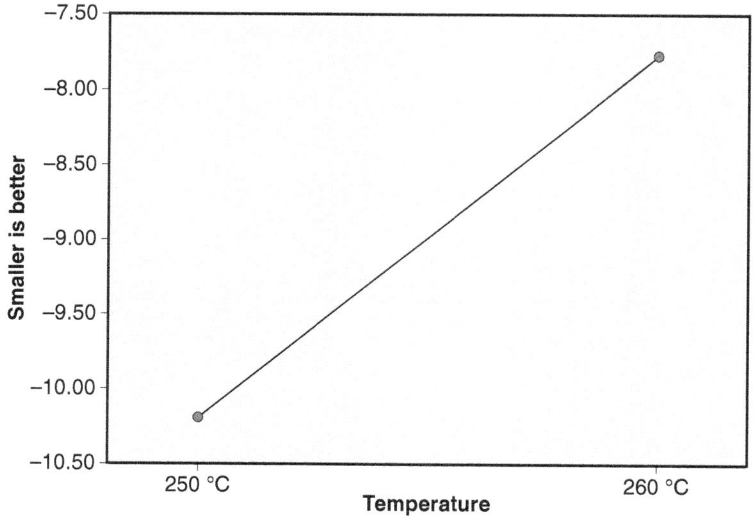

Figure 8.11 Plot of effect 2 (temperature) on opening force.

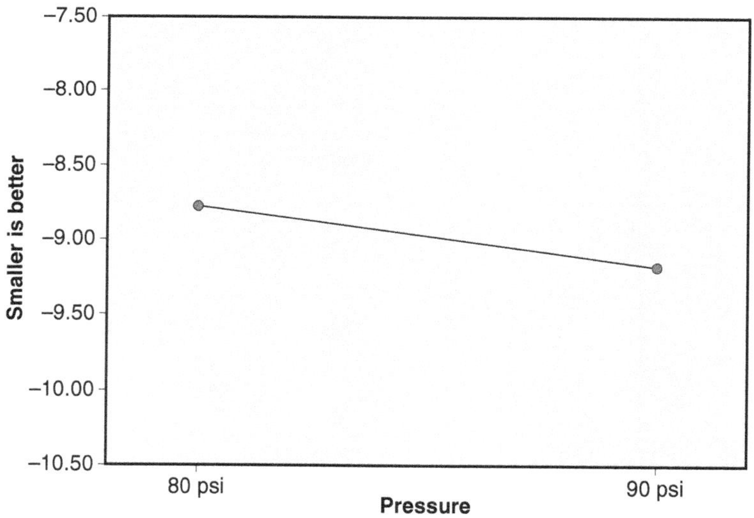

Figure 8.12 Plot of effect 3 (pressure) on opening force.

9. *Model the significant effects, significant interactions, and individual effects from significant interactions.* Recall from Section 8.2 that the goal for a smaller is best S/N ratio is to *maximize* the response.

The coefficients must be calculated so the model can be compiled. The coefficient calculations are simply the average difference of the overall average from a level (1, 2, 3, . . .). Note that it is not necessary to calculate all of the coefficients. All coefficients are calculated for illustrative purposes.

$$C_{1,2,3...} = \bar{Y}_{1,2,3...} - \bar{\bar{Y}}$$

where

$\bar{\bar{Y}}$ = Average run S/N ratio

$\bar{Y}$ = Average treatment level

$\text{Time}_1 = -8.25 - -8.98 = 0.73$

$\text{Time}_2 = -9.72 - -8.98 = -0.74$

$\text{Temp}_1 = -10.19 - -8.98 = -1.21$

$\text{Temp}_2 = -7.77 - -8.98 = 1.21$

$\text{Press}_1 = -8.78 - -8.98 = 0.20$

$\text{Press}_2 = -9.19 - -8.98 = -0.21$

Since our objective is to *minimize* the opening force yield, set factor 1 (time) to 1.25 seconds and factor 2 (temperature) to 260 °C. Factor 3 (pressure) is not significant and can be set solely on the basis of cost, productivity, or convenience. Refer to Figures 8.10, 8.11, and 8.12.

$\hat{Y} = \bar{\bar{Y}} +$ Coefficients of the significant effects, significant interactions, and individual effects from significant interactions

$$\hat{Y} = \bar{\bar{Y}} + 1_1 + 2_2$$

$$\hat{Y} = -8.98 + 0.73 + 1.21 = -7.04$$

To convert the model prediction back to the standard units:

$$Y = \sqrt{\log^{-1}\left(\hat{Y} / -10\right)}$$

where

Y = Predicted result in original units

$\hat{Y}$ = Model prediction

$$Y = \sqrt{\log^{-1}\left(\hat{Y} / -10\right)} = \sqrt{\log^{-1}(-7.04 / -10)} = 2.25$$

With factor 1 (time) set to 1.25 seconds and factor 2 (temperature) set to 260 °C, the pull force is estimated to be 2.25 psi.

At this point it would be wise to verify experimental results by using the model of the experiment to operate the process. All nonsignificant factors can be set solely on the basis of cost, productivity, or convenience. One may also consider opening up the window of the significant factors to see if further process improvement is possible. Remember that most problems are solved by a sequence of studies that move toward an optimum region. This requires that one be aware of physical and safety constraints that may limit further studies. If still not satisfied with the performance, look for other factors.

8.4 TAGUCHI L8 ORTHOGONAL ARRAY EXAMPLE

A heat-treating processor wished to find the parameters to adjust to increase the hardness of a steel alloy. The goal of the experiment is to identify the factor(s) that increase the hardness of the steel alloy. The process has five key variables: preheat, equalize, austenize, temper, and quench (see Table 8.2). The team decided, based on experience and the manufacturer's recommendation, to use the following levels:

	Preheat	**Equalize**	**Austenize**	**Temper**	**Quench**
Level 1	1250 °F	1350 °F	1725 °F	750 °F	150 °F
Level 2	1450 °F	1450 °F	1775 °F	950 °F	200 °F

Because the goal is to increase the steel alloy hardness, *larger is better* is used for a continuous nonnegative number that has a larger is better requirement. The goal is to *maximize* the S/N ratio.

$$S/N = -10\text{Log}_{10}\left[\frac{1}{n}\Sigma\left(\frac{1}{y_1^2}+\frac{1}{y_2^2}+\ldots\frac{1}{y_n^2}\right)\right]$$

where

n = Number of response

y = Response

$$S/N_1 = -10\text{Log}_{10}\left[\frac{1}{n}\Sigma\left(\frac{1}{y_1^2}+\frac{1}{y_2^2}+\frac{1}{y_3^2}\right)\right] = -10\text{Log}_{10}\left[\frac{1}{3}\Sigma\left(\frac{1}{38.0^2}+\frac{1}{36.3^2}+\frac{1}{38.3^2}\right)\right]$$

$$= 31.48$$

$$S/N_2 = -10\text{Log}_{10}\left[\frac{1}{n}\Sigma\left(\frac{1}{y_1^2}+\frac{1}{y_2^2}+\frac{1}{y_3^2}\right)\right] = -10\text{Log}_{10}\left[\frac{1}{3}\Sigma\left(\frac{1}{39.9^2}+\frac{1}{56.2^2}+\frac{1}{35.1^2}\right)\right]$$

$$= 32.32$$

Table 8.2 Steel alloy heat-treating example data.

Run	1 Preheat	2 Equalize	3 Austenize	4 Temper	5 Quench	Y_1	Y_2	Y_3	S/N
1	1	1	1	1	1	38.0	36.3	38.3	31.48
2	1	1	1	2	2	39.9	56.2	35.1	32.32
3	1	2	2	1	1	45.1	65.7	32.1	32.48
4	1	2	2	2	2	41.5	46.8	29.7	31.40
5	2	1	2	1	2	62.1	53.9	61.3	35.38
6	2	1	2	2	1	23.4	19.8	42.9	27.85
7	2	2	1	1	2	33.7	43.0	61.8	32.51
8	2	2	1	2	1	31.4	47.7	31.2	30.82
ΣY_1	127.68	127.03	127.13	131.85	122.63				
ΣY_2	126.56	127.21	127.11	122.39	131.61				$\overline{\overline{Y}}$ 31.88
$\overline{Y}_1$	31.92	31.76	31.78	32.96	30.66				
$\overline{Y}_2$	31.64	31.80	31.78	30.60	32.90				
C_1	0.04	–0.12	–0.10	1.08	–1.22				
C_2	–0.24	–0.08	–0.10	–1.28	1.02				
Abs effect	0.28	0.04	0.00	2.36	2.24				

$$S/N_3 = -10\text{Log}_{10}\left[\frac{1}{n}\Sigma\left(\frac{1}{y_1^2}+\frac{1}{y_2^2}+\frac{1}{y_3^2}\right)\right] = -10\text{Log}_{10}\left[\frac{1}{3}\Sigma\left(\frac{1}{45.1^2}+\frac{1}{65.7^2}+\frac{1}{32.1^2}\right)\right]$$

$$= 32.48$$

$$S/N_4 = -10\text{Log}_{10}\left[\frac{1}{n}\Sigma\left(\frac{1}{y_1^2}+\frac{1}{y_2^2}+\frac{1}{y_3^2}\right)\right] = -10\text{Log}_{10}\left[\frac{1}{3}\Sigma\left(\frac{1}{41.5^2}+\frac{1}{46.8^2}+\frac{1}{29.7^2}\right)\right]$$

$$= 31.40$$

$$S/N_5 = -10\text{Log}_{10}\left[\frac{1}{n}\Sigma\left(\frac{1}{y_1^2}+\frac{1}{y_2^2}+\frac{1}{y_3^2}\right)\right] = -10\text{Log}_{10}\left[\frac{1}{3}\Sigma\left(\frac{1}{62.1^2}+\frac{1}{53.9^2}+\frac{1}{61.3^2}\right)\right]$$

$$= 35.38$$

$$S/N_6 = -10\text{Log}_{10}\left[\frac{1}{n}\Sigma\left(\frac{1}{y_1^2}+\frac{1}{y_2^2}+\frac{1}{y_3^2}\right)\right] = -10\text{Log}_{10}\left[\frac{1}{3}\Sigma\left(\frac{1}{23.4^2}+\frac{1}{19.8^2}+\frac{1}{42.9^2}\right)\right]$$

$$= 27.85$$

$$S/N_7 = -10\text{Log}_{10}\left[\frac{1}{n}\Sigma\left(\frac{1}{y_1^2}+\frac{1}{y_2^2}+\frac{1}{y_3^2}\right)\right] = -10\text{Log}_{10}\left[\frac{1}{3}\Sigma\left(\frac{1}{33.7^2}+\frac{1}{43.0^2}+\frac{1}{61.8^2}\right)\right]$$

$$= 32.51$$

$$S/N_8 = -10\text{Log}_{10}\left[\frac{1}{n}\Sigma\left(\frac{1}{y_1^2}+\frac{1}{y_2^2}+\frac{1}{y_3^2}\right)\right] = -10\text{Log}_{10}\left[\frac{1}{3}\Sigma\left(\frac{1}{31.4^2}+\frac{1}{47.7^2}+\frac{1}{31.2^2}\right)\right]$$

$$= 30.82$$

Nine Steps for Analysis of Effects

1. *Calculate the absolute values of the effects:*

$$|E(1)| = |\bar{Y}_1| - |\bar{Y}_2| = \frac{31.48+32.32+32.48+31.40}{4} - \frac{35.38+27.85+32.51+30.82}{4}$$

$$= 0.28$$

$$|E(2)| = |\bar{Y}_1| - |\bar{Y}_2| = \frac{31.48+32.32+35.38+27.85}{4} - \frac{32.48+31.40+32.51+30.82}{4}$$

$$= 0.04$$

$$|E(3)| = |\overline{Y}_1| - |\overline{Y}_2| = \frac{31.48+32.32+32.51+30.82}{4} - \frac{32.48+31.40+35.38+27.85}{4}$$

$$= 0.00$$

$$|E(4)| = |\overline{Y}_1| - |\overline{Y}_2| = \frac{31.48+32.48+35.38+32.51}{4} - \frac{32.32+31.40+27.85+30.82}{4}$$

$$= 2.36$$

$$|E(5)| = |\overline{Y}_1| - |\overline{Y}_2| = \frac{31.48+32.48+27.85+30.82}{4} - \frac{32.32+31.40+35.38+32.51}{4}$$

$$= 2.24$$

Note: The minus signs have been removed for simplicity.

2. *Create a Pareto chart of the absolute values of the effects.* The Pareto chart in Figure 8.13 indicates that factor 4 (temper) and factor 5 (quench) may be significant.

3. *Calculate the standard deviation of the experiment, s_e.* This step does not apply when using S/N ratios.

4. *Calculate the standard deviation of the effects, s_{Eff}.* This step does not apply when using S/N ratios.

5. *Determine the* t-*statistic.* This step does not apply when using S/N ratios.

6. *Calculate the decision limits.* This step does not apply when using S/N ratios.

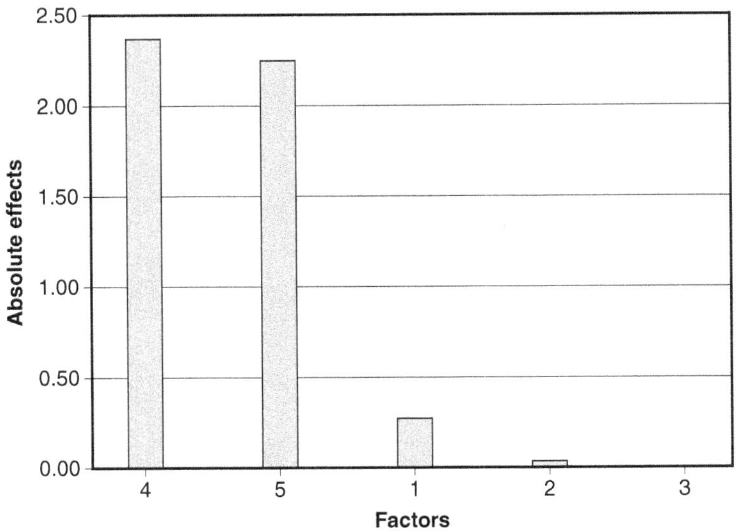

Figure 8.13 Pareto chart of the absolute values of the effects.

7. *Determine the significant effects.* An analysis of Figure 8.13 and 8.14 indicates that factor 4 (temper) and factor 5 (quench) are significant.

8. *Graph the main effects and any significant interactions* (Figures 8.15, 8.16, 8.17, 8.18, 8.19).

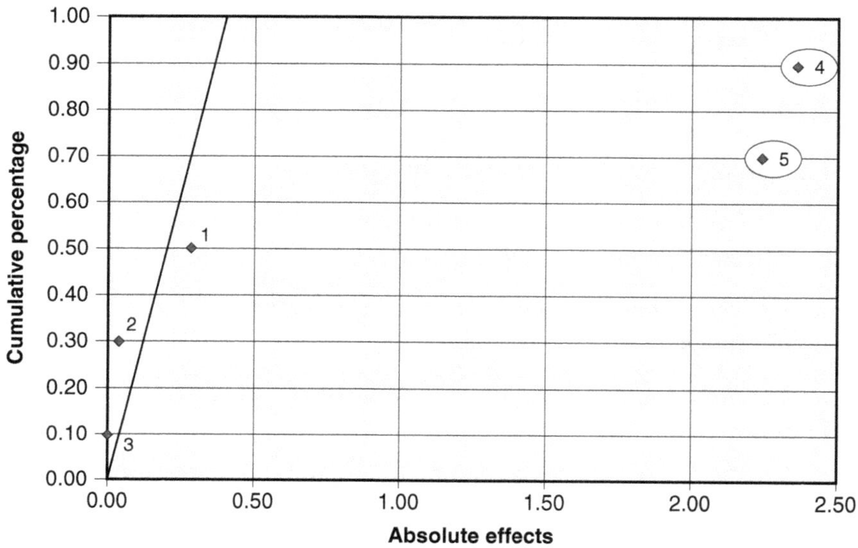

Figure 8.14 Half-normal plot of the absolute effects.

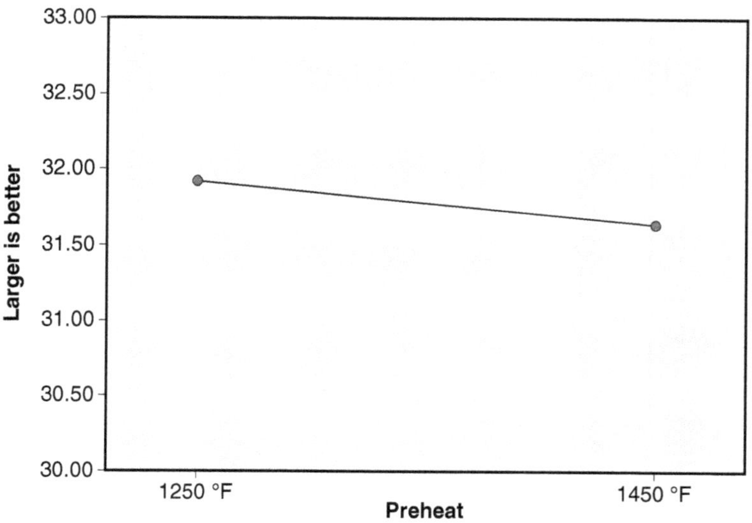

Figure 8.15 Plot of effect 1 (preheat) on hardness.

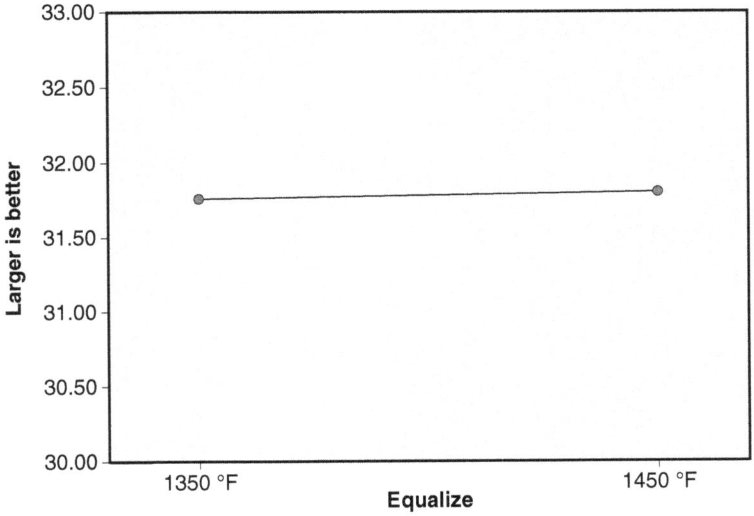

Figure 8.16 Plot of effect 2 (equalize) on hardness.

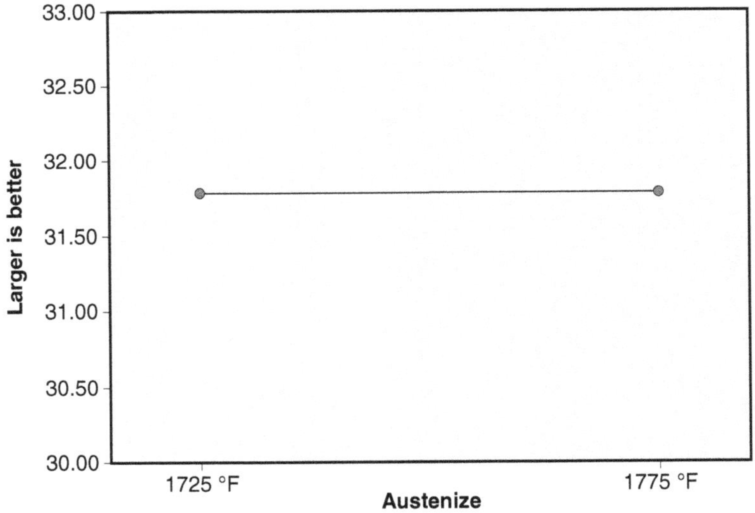

Figure 8.17 Plot of effect 3 (austenize) on hardness.

9. *Model the significant effects, significant interactions, and individual effects from significant interactions.* Recall from Section 8.2 that the goal for a larger is best S/N ratio is to *maximize* the response.

The coefficients must be calculated so the model can be compiled. The coefficient calculations are simply the average difference of the overall average from a level (1, 2, 3, . . .). Note that it is not necessary to calculate all of the coefficients. All coefficients are calculated for illustrative purposes.

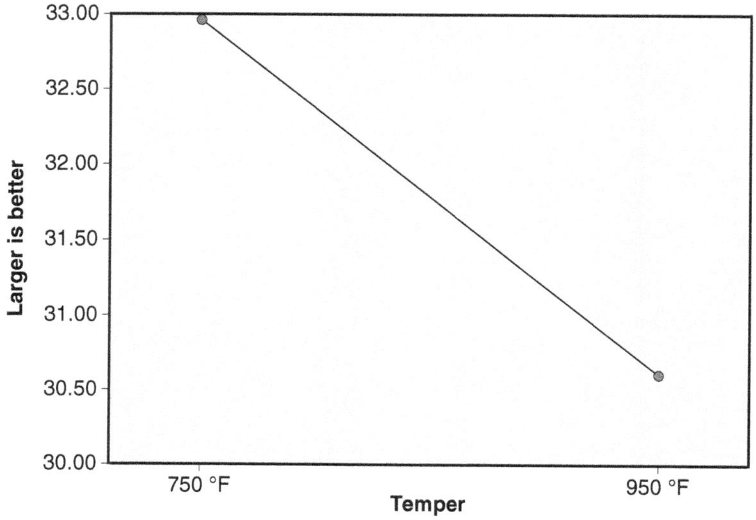

Figure 8.18 Plot of effect 4 (temper) on hardness.

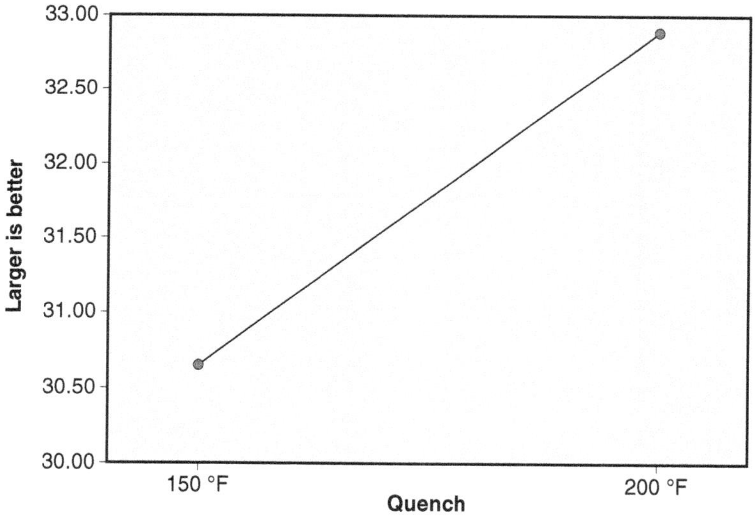

Figure 8.19 Plot of effect 5 (quench) on hardness.

$$C_{1,2,3\ldots} = \bar{Y}_{1,2,3\ldots} - \bar{\bar{Y}}$$

where

$\bar{\bar{Y}}$ = Average run S/N ratio

$\bar{Y}$ = Average treatment level

$\text{Preheat}_1 = 31.92 - 31.88 = 0.04$

$\text{Preheat}_2 = 31.64 - 31.88 = -0.24$

$\text{Equalize}_1 = 31.76 - 31.88 = -0.12$

$\text{Equalize}_2 = 31.80 - 31.88 = -0.08$

$\text{Austenize}_1 = 31.78 - 31.88 = -0.10$

$\text{Austenize}_2 = 31.78 - 31.88 = -0.10$

$\text{Temper}_1 = 32.96 - 31.88 = 1.08$

$\text{Temper}_2 = 30.60 - 31.88 = -1.28$

$\text{Quench}_1 = 30.66 - 31.88 = -1.22$

$\text{Quench}_2 = 32.90 - 31.88 = 1.02$

Since our objective is to *maximize* the steel alloy hardness, set factor 4 (temper) to 750 °F and factor 5 (quench) to 200 °F. Factor 1 (preheat), factor 2 (equalize), and factor 3 (austenize) are not significant and can be set solely on the basis of cost, productivity, or convenience. Refer to Figures 8.18 and 8.19.

$\hat{Y} = \bar{\bar{Y}} +$ Coefficients of the significant effects, significant interactions, and individual effects from significant interactions

$$\hat{Y} = \bar{\bar{Y}} + 4_1 + 5_2$$

$$\hat{Y} = 31.88 + 1.08 + 1.02 = 33.98$$

To convert the model prediction back to the standard units:

$$Y = \sqrt{\log^{-1}\left(\hat{Y}/10\right)}$$

where

Y = Predicted result in original units

$\hat{Y}$ = Model prediction

$$Y = \sqrt{\log^{-1}\left(\hat{Y}/10\right)} = \sqrt{\log^{-1}(33.98/10)} = 50.00$$

With factor 4 (temper) set to 750 °F and factor 5 (quench) set to 200 °F, the hardness is estimated to be 50 HRC.

At this point it would be wise to verify experimental results by using the model of the experiment to operate the process. All nonsignificant factors can be set solely on the basis of cost, productivity, or convenience. One may also consider opening up the window of the significant factors to see if further process improvement is possible.

Remember that most problems are solved by a sequence of studies that move toward an optimum region. This requires that one be aware of physical and safety constraints that may limit further studies. If still not satisfied with the performance, look for other factors.

8.5 TAGUCHI L9 ORTHOGONAL ARRAY EXAMPLE

A plastics processor wishes to find the parameters that will allow them to consistently produce parts to the required nominal specification. The goal of the experiment is to identify the factor(s) that allow for the manufacture of the product to the nominal specification. The process has four key variables: temperature, time, pressure, and polymer (see Table 8.3). The team decided, based on experience and the manufacturer's recommendation, to use the following levels:

	Temp	**Time**	**Pressure**	**Polymer**
Level 1	125	80	2	0.5
Level 2	150	85	4	1.0
Level 3	200	90	6	1.5

Because the goal is to produce parts to the nominal specification, *nominal is best* is used for a continuous nonnegative number that has a specified nominal or target value. The goal is to *maximize* the S/N ratio. Please note that the variance or standard deviation cannot be zero.

$$\text{S/N} = 10\text{Log}_{10}\left(\bar{Y}^2 / s^2\right)$$

where

$\bar{Y}$ = Mean

s = Standard deviation

$$\text{S/N}_1 = 10\text{Log}_{10}\left(\bar{Y}^2 / s^2\right) = 10\text{Log}_{10}\left(80.30^2 / 9.13^2\right) = 18.88$$

$$\text{S/N}_2 = 10\text{Log}_{10}\left(\bar{Y}^2 / s^2\right) = 10\text{Log}_{10}\left(69.33^2 / 5.30^2\right) = 22.33$$

$$\text{S/N}_3 = 10\text{Log}_{10}\left(\bar{Y}^2 / s^2\right) = 10\text{Log}_{10}\left(51.67^2 / 6.01^2\right) = 18.69$$

$$\text{S/N}_4 = 10\text{Log}_{10}\left(\bar{Y}^2 / s^2\right) = 10\text{Log}_{10}\left(74.10^2 / 9.46^2\right) = 17.88$$

Table 8.3 Plastic processing example data.

Run	1 Temp	2 Time	3 Pressure	4 Polymer	Y_1	Y_2	Y_3	$\bar{Y}$	s	S/N
1	1	1	1	1	88.2	82.4	70.3	80.30	9.13	18.88
2	1	2	2	2	74.7	69.2	64.1	69.33	5.30	22.33
3	1	3	3	3	56.4	53.7	44.9	51.67	6.01	18.69
4	2	1	2	3	80.2	78.9	63.2	74.10	9.46	17.88
5	2	2	3	1	77.4	76.2	53.9	69.17	13.23	14.37
6	2	3	1	2	88.9	88.1	82.9	86.63	3.26	28.48
7	3	1	3	2	64.3	61.9	56.1	60.77	4.22	23.17
8	3	2	1	3	98.6	92.6	88.8	93.33	4.94	25.53
9	3	3	2	1	75.9	73.4	62.8	70.70	6.95	20.11
ΣY_1	59.90	59.93	72.89	53.36						
ΣY_2	60.73	62.23	60.32	73.98						$\bar{\bar{Y}}$ 21.07
ΣY_3	68.81	67.28	56.23	62.10						
$\bar{Y}_1$	19.97	19.98	24.30	17.79						
$\bar{Y}_2$	20.24	20.74	20.11	24.66						
$\bar{Y}_3$	22.94	22.43	18.74	20.70						
C_1	−1.10	−1.09	3.23	−3.28						
C_2	−0.83	−0.33	−0.96	3.59						
C_3	1.87	1.36	−2.33	−0.37						
Abs effect	2.97	2.45	5.56	6.87						

$$S/N_5 = 10\text{Log}_{10}\left(\bar{Y}^2 / s^2\right) = 10\text{Log}_{10}\left(69.17^2 / 13.23^2\right) = 14.37$$

$$S/N_6 = 10\text{Log}_{10}\left(\bar{Y}^2 / s^2\right) = 10\text{Log}_{10}\left(86.63^2 / 3.26^2\right) = 28.48$$

$$S/N_7 = 10\text{Log}_{10}\left(\bar{Y}^2 / s^2\right) = 10\text{Log}_{10}\left(60.77^2 / 4.22^2\right) = 23.17$$

$$S/N_8 = 10\text{Log}_{10}\left(\bar{Y}^2 / s^2\right) = 10\text{Log}_{10}\left(93.33^2 / 4.94^2\right) = 25.53$$

$$S/N_9 = 10\text{Log}_{10}\left(\bar{Y}^2 / s^2\right) = 10\text{Log}_{10}\left(70.70^2 / 6.95^2\right) = 20.11$$

Nine Steps for Analysis of Effects

1. *Calculate the absolute values of the effects*:

$$|E(1)| = \max\left|\bar{Y}_{1,2,3}\right| - \min\left|\bar{Y}_{1,2,3}\right| = 22.94 - 19.97 = 2.97$$

$$|E(2)| = \max\left|\bar{Y}_{1,2,3}\right| - \min\left|\bar{Y}_{1,2,3}\right| = 22.43 - 19.98 = 2.45$$

$$|E(3)| = \max\left|\bar{Y}_{1,2,3}\right| - \min\left|\bar{Y}_{1,2,3}\right| = 24.30 - 18.74 = 5.56$$

$$|E(4)| = \max\left|\bar{Y}_{1,2,3}\right| - \min\left|\bar{Y}_{1,2,3}\right| = 24.66 - 17.79 = 6.87$$

Note: The values from Table 8.3 have been used for simplicity.

2. *Create a Pareto chart of the absolute values of the effects.* The Pareto chart in Figure 8.20 indicates that factor 4 (polymer) and factor 3 (pressure) may be significant.

3. *Calculate the standard deviation of the experiment*, s_e. This step does not apply when using S/N ratios.

4. *Calculate the standard deviation of the effects*, s_{Eff}. This step does not apply when using S/N ratios.

5. *Determine the* t-*statistic.* This step does not apply when using S/N ratios.

6. *Calculate the decision limits.* This step does not apply when using S/N ratios.

7. *Determine the significant effects.* An analysis of Figure 8.20 indicates that factor 4 (polymer) and factor 3 (pressure) may be significant.

A half-normal probability plot may not be appropriate for this example because the number of effects is small (four-factor experiment). The results using this method may be erratic. The Pareto chart and factor plots should be evaluated to ensure an appropriate decision is made.

8. *Graph the main effects and any significant interactions* (Figures 8.21, 8.22, 8.23, and 8.24).

9. *Model the significant effects, significant interactions, and individual effects from significant interactions.* Recall from Section 8.2 that the goal for nominal is best S/N ratio is to *maximize* the response.

The coefficients must be calculated so the model can be compiled. The coefficient calculations are simply the average difference of the overall average from a level (1, 2,

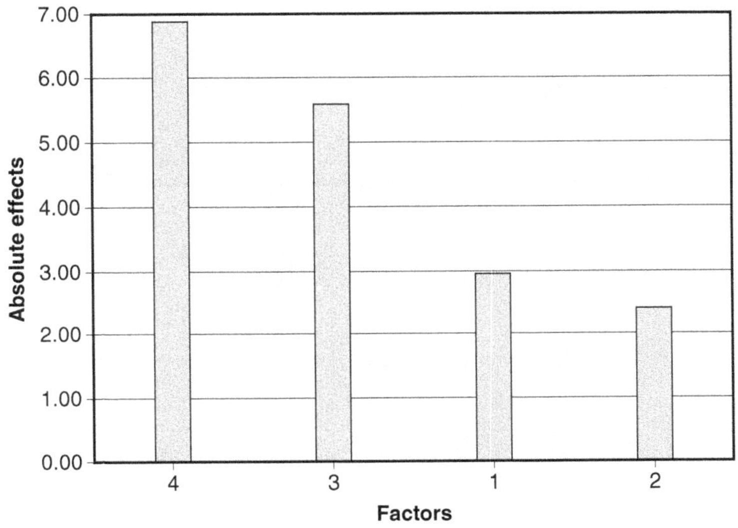

Figure 8.20 Pareto chart of the absolute values of the effects.

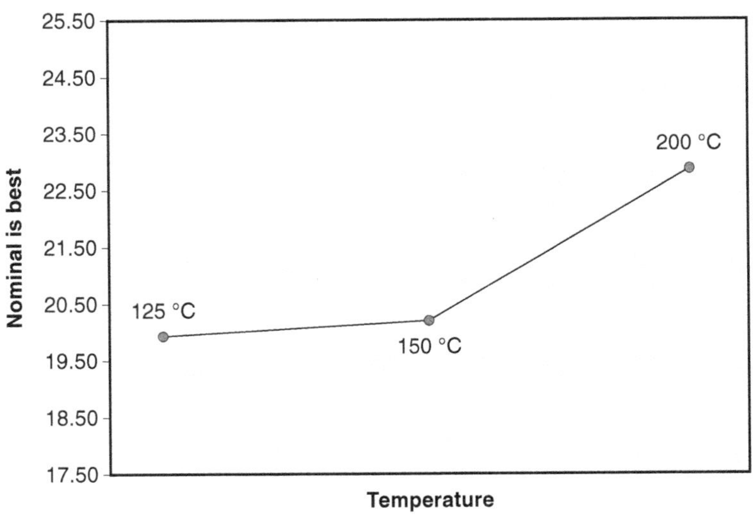

Figure 8.21 Plot of effect 1 (temperature) on the nominal specification.

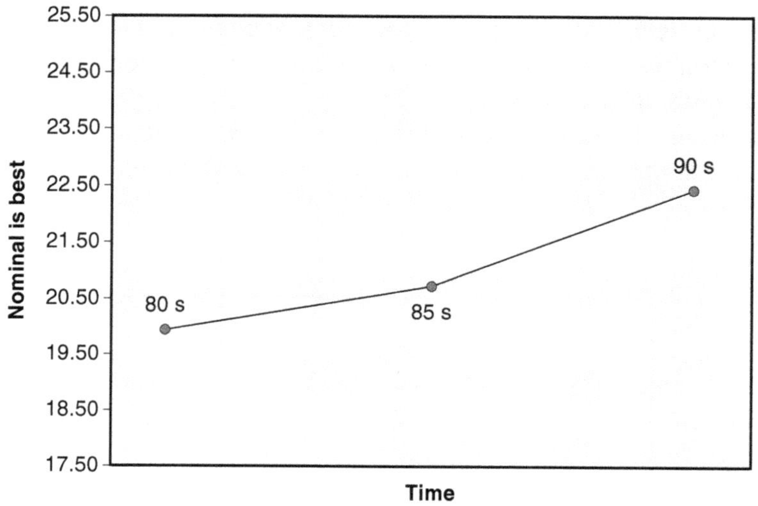

Figure 8.22 Plot of effect 2 (time) on the nominal specification.

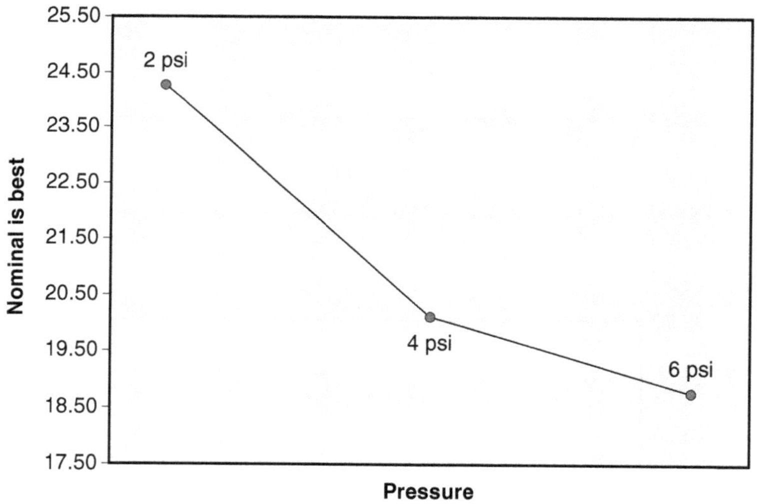

Figure 8.23 Plot of effect 3 (pressure) on the nominal specification.

3, . . .). Note that it is not necessary to calculate all of the coefficients. All coefficients are calculated for illustrative purposes.

$$C_{1,2,3\ldots} = \bar{Y}_{1,2,3\ldots} - \bar{\bar{Y}}$$

where

$\bar{\bar{Y}}$ = Average run S/N ratio

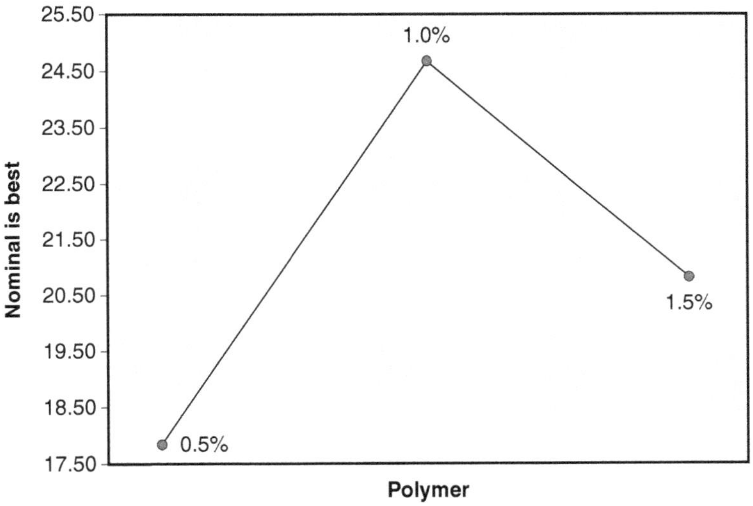

Figure 8.24 Plot of effect 4 (polymer) on the nominal specification.

$\bar{Y}$ = Average treatment level

$Temperature_1 = 19.97 - 21.07 = -1.10$

$Temperature_2 = 20.24 - 21.07 = -0.83$

$Temperature_3 = 22.94 - 21.07 = -1.87$

$Time_1 = 19.98 - 21.07 = -1.09$

$Time_2 = 20.74 - 21.07 = -0.33$

$Time_3 = 22.43 - 21.07 = -1.36$

$Pressure_1 = 24.30 - 21.07 = 3.23$

$Pressure_2 = 20.11 - 21.07 = -0.96$

$Pressure_3 = 18.74 - 21.07 = -2.33$

$Polymer_1 = 17.79 - 21.07 = -3.28$

$Polymer_2 = 24.66 - 21.07 = 3.59$

$Polymer_3 = 20.70 - 21.07 = -0.37$

Since our objective is to produce product to the nominal specification, set factor 3 (pressure) to 2 psi and set factor 4 (polymer) to 1%. Factor 1 (temperature) and factor 2 (time) are not significant and can be set solely on the basis of cost, productivity, or convenience. Refer to Figures 8.23 and 8.24.

$$\hat{Y} = \bar{\bar{Y}} + \text{Coefficients of the significant effects, significant interactions, and individual effects from significant interactions}$$

$$\hat{Y} = \bar{\bar{Y}} + 3_1 + 4_2$$

$$\hat{Y} = 21.07 + 3.23 + 3.59 = 27.89$$

At this point it would be wise to verify experimental results by using the model of the experiment to operate the process. All nonsignificant factors can be set solely on the basis of cost, productivity, or convenience. One may also consider opening up the window of the significant factors to see if further process improvement is possible. Remember that most problems are solved by a sequence of studies that move toward an optimum region. This requires that one be aware of physical and safety constraints that may limit further studies. If still not satisfied with the performance, look for other factors.

9

Mixture Designs

Mixture design experiments represent a special type of experiment in which the product is made up of several components or ingredients. This type of experiment is useful in situations involving formulations or mixtures. The output response is related to the proportions of the different components of the formulation or ingredients in the mixture. Mixture experiments depend on the proportions of the formulation or ingredients. The amounts of each component must add up to a total of 1 or 100%. Mixture designs are the most useful type of experiment in determining the best composition of a product.

Figure 9.1 depicts a typical three-component mixture design consisting of components (1, 2, 3). The vertices in the simplex region represent a 100% composition of each individual component. The midpoints on each side correspond to 50% mixtures of the components of each connecting vertex, and the central point represents a mixture with equal amounts of each of the components. Moving from a linear to a quadratic, reduced cubic, or full cubic model (see Figure 9.2) represents adding more experimental points to the design, allowing it to start with the simplest, least time-consuming experiment, and adding additional points and complexity as needed.

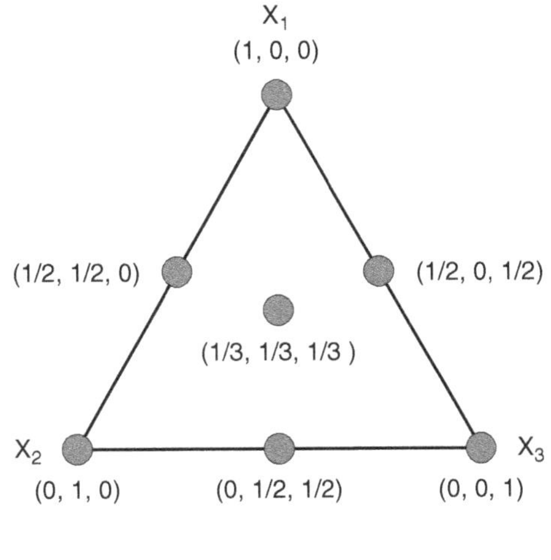

Figure 9.1 Three-component reduced cubic mixture design.

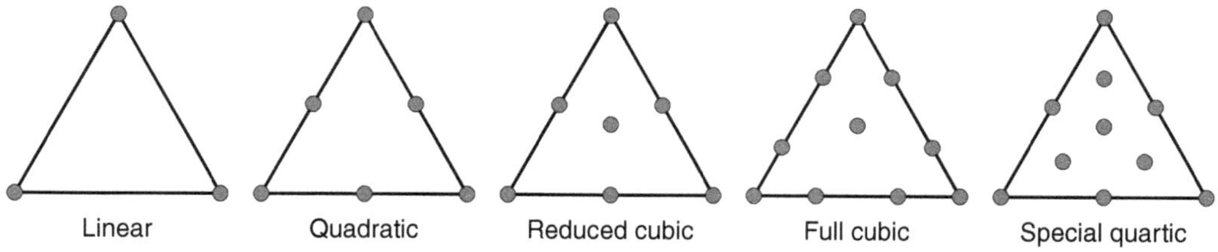

Figure 9.2 Linear, quadratic, reduced cubic, full cubic, and special quartic mixture designs.

Four standard models are used for fitting mixture data (Cornell 1990):

- Linear model:

$$\hat{Y} = \beta_1 x_1 + \beta_2 x_2 + \beta_3 x_3$$

- Quadratic model:

$$\hat{Y} = \beta_1 x_1 + \beta_2 x_2 + \beta_3 x_3 + \beta_{12} x_1 x_2 + \beta_{13} x_1 x_3 + \beta_{23} x_2 x_3$$

- Reduced cubic model:

$$\hat{Y} = \beta_1 x_1 + \beta_2 x_2 + \beta_3 x_3 + \beta_{12} x_1 x_2 + \beta_{13} x_1 x_3 + \beta_{23} x_2 x_3 + \beta_{123} x_1 x_2 x_3$$

- Full cubic model:

$$\hat{Y} = \beta_1 x_1 + \beta_2 x_2 + \beta_3 x_3 + \beta_{12} x_1 x_2 + \beta_{13} x_1 x_3 + \beta_{23} x_2 x_3$$
$$+ \beta_{12} x_1 x_2 (x_1 - x_2) + \beta_{13} x_1 x_3 (x_1 - x_3) + \beta_{23} x_2 x_3 (x_2 - x_3) + \beta_{123} x_1 x_2 x_3$$

where

$x_{1,2,3,\text{ etc.}}$ = Factor

$\beta_{1,2,3,\text{ etc.}}$ = Coefficient

To calculate the coefficients (β), we use the following series of equations:

$$\beta_1 = \bar{Y}x_1$$
$$\beta_2 = \bar{Y}x_2$$
$$\beta_3 = \bar{Y}x_3$$
$$\beta_{12} = 4(\bar{Y}x_1x_2) - 2(\bar{Y}x_1 + \bar{Y}x_2)$$
$$\beta_{13} = 4(\bar{Y}x_1x_3) - 2(\bar{Y}x_1 + \bar{Y}x_3)$$
$$\beta_{23} = 4(\bar{Y}x_2x_3) - 2(\bar{Y}x_2 + \bar{Y}x_3)$$
$$\beta_{123} = 27(\bar{Y}x_1x_2x_3) - 12(\bar{Y}x_{12} + \bar{Y}x_{13} + \bar{Y}x_{23}) + 3(\bar{Y}x_1 + \bar{Y}x_2 + \bar{Y}x_3)$$

The analysis procedure for mixture designs is as follows:

1. Calculate the coefficients (β)
2. Construct the model of the experiment

It should be noted that there are additional calculations available to check the goodness of fit of the various models. Please consult Cornell (1990) for additional information.

9.1 THREE-FACTOR SIMPLEX DESIGN EXAMPLE

A type of blown film used for packaging consists of three primary components: nylon, EVOH, and PET. The manufacturer's marketing team wishes to discover which combination of these materials produces the best film characteristics. The data for the experiment are shown in Table 9.1. The response nomenclature is depicted in Figure 9.3.

Table 9.1 Blown film development example data.

	X_1 Nylon	X_2 EVOH	X_3 PET	$\bar{Y}$	Coef
X_1	1	0	0	86.1	86.1
X_2	0	1	0	80.9	80.9
X_3	0	0	1	78.6	78.6
$X_1 X_2$	0.5	0.5	0	81.7	–7.2
$X_1 X_3$	0.5	0	0.5	81.4	–3.8
$X_2 X_3$	0	0.5	0.5	81.4	6.6

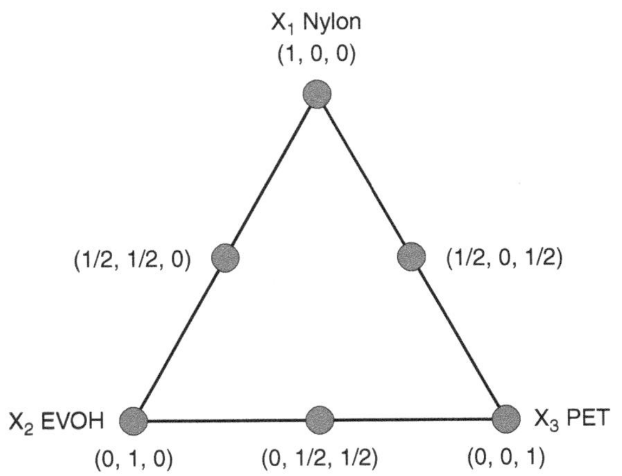

Figure 9.3 Three-component quadratic mixture design for the blown film example.

The analysis procedure is as follows:

1. *Calculate the coefficients* (β):

$$\beta_1 = \overline{Y}x_1 = 86.1$$

$$\beta_2 = \overline{Y}x_2 = 80.9$$

$$\beta_3 = \overline{Y}x_3 = 78.6$$

$$\beta_{12} = 4(\overline{Y}x_1x_2) - 2(\overline{Y}x_1 + \overline{Y}x_2) = 4(81.7) - 2(86.1 + 80.9) = -7.2$$

$$\beta_{13} = 4(\overline{Y}x_1x_3) - 2(\overline{Y}x_1 + \overline{Y}x_3) = 4(81.4) - 2(86.1 + 78.6) = -3.8$$

$$\beta_{23} = 4(\overline{Y}x_2x_3) - 2(\overline{Y}x_2 + \overline{Y}x_3) = 4(81.4) - 2(80.9 + 78.6) = 6.6$$

2. *Construct the model of the experiment*:

   - Quadratic model:

$$\hat{Y} = \beta_1 x_1 + \beta_2 x_2 + \beta_3 x_3 + \beta_{12} x_1 x_2 + \beta_{13} x_1 x_3 + \beta_{23} x_2 x_3$$

   - With the coefficients placed, the model becomes:

$$\hat{Y} = 86.1x_1 + 80.9x_2 + 78.6x_3 + -7.2x_1x_2 + -3.8x_1x_3 + 6.6x_2x_3$$

Because coefficients β_{12} and β_{13} are negative, they are antagonistic. Because β_{23} is positive, it is synergistic. *Synergism* is the positive excess of a binary mixture. *Antagonism* is the opposite of synergism.

We now wish to use the model to make some additional predictions. The engineers wish to try four additional combinations as shown in Table 9.2.

For trial 1:

$$\hat{Y} = 86.1(0.34) + 80.9(0.33) + 78.6(0.33) + -7.2(0.34 \times 0.33) \\ + -3.8(0.34 \times 0.33) + 6.6(0.33 \times 0.33) = 81.4$$

Table 9.2 Blown film development example—additional data.

	X_1 Nylon	X_2 EVOH	X_3 PET
1	0.34	0.33	0.33
2	0.72	0.14	0.14
3	0.14	0.57	0.29
4	0.14	0.29	0.57

For trial 2:

$$\hat{Y} = 86.1(0.72) + 80.9(0.14) + 78.6(0.14) + -7.2(0.72 \times 0.14) \\ + -3.8(0.72 \times 0.14) + 6.6(0.14 \times 0.14) = 85.9$$

For trial 3:

$$\hat{Y} = 86.1(0.14) + 80.9(0.57) + 78.6(0.29) + -7.2(0.14 \times 0.57) \\ + -3.8(0.14 \times 0.29) + 6.6(0.57 \times 0.29) = 80.3$$

For trial 4:

$$\hat{Y} = 86.1(0.14) + 80.9(0.29) + 78.6(0.57) + -7.2(0.14 \times 0.29) \\ + -3.8(0.14 \times 0.57) + 6.6(0.29 \times 0.57) = 79.8$$

Based on the model prediction, the most favorable combination is trial 2 using X_1 nylon (0.72), X_2 EVOH(0.14), and X_3 PET(0.14), yielding the best result of 85.9.

At this point it would be wise to verify experimental results by using the model of the experiment to operate the process. One may also consider additional percentages to see if further process improvement is possible. Remember that most problems are solved by a sequence of studies that move toward an optimum region. This requires that one be aware of physical and safety constraints that may limit further studies. If still not satisfied with the performance, look for other favorable percentages.

10
Procedural Considerations

10.1 COMMON PROBLEMS AND QUESTIONS

Frequently asked questions and technical points that need to be repeated include:

1. *If one cannot do both, should the experimenter reflect or replicate a screening design?* Reflection is always preferred except when there is concern about special cause variation during the DOE. In such cases, replication is preferred. The rationale is that assessment of significance can be done quite well with a Pareto chart, partially overcoming the lack of a formal test of significance that is possible with replication. Reflection provides clarity of results on main effects. However, if special causes abound, one will have difficulty drawing accurate conclusions without replication. Assessment of interaction is very risky in the face of random special causes.

2. *What if many of the effects are significant?* The experimenter should carefully review the randomization pattern used while performing the experiment. Was the setup for each trial really randomly replicated? If not, the responses are repeats, not replicates. Replication assumes that each trial is an independent and random performance of the process, specifically including any process setup. If not, the experimenter has repeats that may be neither independent nor random. The estimate of s_e using repeats will be much smaller than the actual inherent variation of the process. This will make most effects appear significant, when in reality they have not been randomly replicated to properly estimate the inherent variation.

3. *What if several factors are significant in the variation analysis?* Variation analysis is very susceptible to outliers or special causes that create major distortions of the variances of each run. Check to see if one or two estimated variances are substantially larger than the others. Review the raw data for outliers, that is, unusual values compared to the other observations. Apparent outliers should be investigated and, if determined to be outliers, removed or replaced with the cell average.

4. *Can one analyze an experiment with missing data?* The simplest thing to do is to replace a missing observation with the grand average if the experiment is unreplicated, or with the average of the other observations for that run if replicated. This minimizes the loss of information. If more than one observation is lost in an unreplicated DOE, computer analysis may be the only way to salvage the analysis depending on the type of DOE performed.

5. *Does one have to use the maximum number of factors for a design?* No. While all runs of a design must be carried out, one may use less than the maximum number of factors. For example, in a 16-run Plackett-Burman design, there is nothing wrong with having fewer than fifteen factors. The unused factors are often referred to as *dummy factors.* The columns of the unused factors represent the family of interactions that would ordinarily have been confounded with those factors had they been used. Interactions with the unused factor(s) do not exist.

It is to the experimenter's advantage to study as many factors as possible in an experiment. Don't use fewer than the maximum number of factors of a design if there are any reasonable factors that could be added to the factor list.

6. *Should one use Plackett-Burman designs or fractional factorial designs?* This question must be answered in two parts.

a. Geometric designs, whether Plackett-Burman or fractional factorials, are equivalent as long as the maximum number of factors is used in Plackett-Burman designs. For example, an eight-run Plackett-Burman design with seven factors is equivalent to an eight-run fractional factorial design with seven factors. However, as fewer than seven factors are used, there comes a point where fractional factorials provide better optimization of the confounding of two-factor interactions. That is why eight-run fractional factorial designs are provided in Appendix I for four and five factors. This same point holds for larger geometric designs, for example, the 16-run or 32-run designs. The fewer the factors (below the maximum possible), the more advantageous the fractional factorial designs will be in minimizing confounding.

b. Nongeometric designs are uniquely Plackett-Burman designs, with the disadvantages discussed in Chapter 7. Nongeometric designs incur more risks than geometric designs in accurately estimating main effects. However, the experimenter should use a nongeometric design when it fits his or her needs. For instance, if there are 10 or 11 factors to be studied, one should not use a 16-run geometric design just to avoid a 12-run nongeometric design. There is, as usual, an exception: if the process is known or suspected to be highly interactive, it would be better to have a geometric design in order to provide at least a grouping of interactions.

Note that reflection is essential to provide clean estimates of main effects in either case! Screening designs are not meant to provide significant information on interactions.

7. *What is blocking and how is it used?* There are times when the experiment cannot be completely randomized. For instance, suppose there is not enough raw material to complete the experiment with one lot, and the experimenter fears that lot differences could create a distortion in the responses. This could be handled by letting raw material lot be a factor and randomizing within each lot. Another example would be an experiment that could not be completed in one day, and a time or weather (temperature/

humidity/barometric pressure) effect was considered a possibility. Again, let one factor be *day*, with day one being the high level and day two being the low level; randomize all the trials within each day.

8. *What if one can replicate a few runs but cannot afford to replicate all of them?*

a. Randomly select the runs to be replicated and add these trials to the design.

b. After the data are available, calculate variances for each replicated run.

c. Calculate the average variance; this is the estimate of s_e. The degrees of freedom are:

$$(\text{Number of replicates} - 1) \times (\text{Number of replicated runs})$$

d. Use the average of the replicates in the analysis table and proceed as usual to estimate effects.

e. Use the calculated s^2 for significance testing. Do not perform the variation analysis.

9. *Should one run one big screening experiment and quit?* No! Experimentation is usually a series of experiments, one building from another. The experiments are generally of diminishing size as the studies move from screening to optimization. While we all get lucky sometimes, it is not typical to get all of your answers in the first experiment.

10. *Must one randomize all trials?* No, although it would be the safest strategy. If there is concern about the feasibility of a run, it would be prudent to carry it out first in case the levels must be redefined. It is best to learn this early, before the study has really gotten under way.

In other experiments a physical constraint may prevent randomization. Assume a furnace temperature must be set and then trials run for that temperature. (In other words, one cannot bounce back and forth.) Be aware that this flinching on the randomization could lead to confounding temperature with an unknown blocking factor.

There are many other situations that may arise. The main rules to remember in dealing with peculiar results are:

- Use common sense
- Look at the data
- Use process knowledge

10.2 REVIEW OF THE BASICS IN MANAGING A DOE

Most people assume that the analytical techniques are the most difficult part of applying DOE. This is not true. The most difficult aspects of any DOE are the planning of the experiment and the actual running of the experiment. Following is a review of the basics of application of DOE:

1. Identify the process to be studied.
2. Precisely define the objectives of the study.
3. Assemble a small team knowledgeable in the process being addressed.
4. Review all pertinent data and knowledge relevant to the process.
5. Brainstorm to generate potential factors. Be creative, and do not accept existing theories without data.
6. Distill from this list those factors that can be controlled and whose cost to the experiment can be tolerated. (The more, the better.) Be sure to include those factors that are cheap, quick, and easy to study.
7. Set levels boldly without being careless. One needs levels as wide as reasonable to force effects to show themselves. Yet, process knowledge must be used to avoid conditions that are either not feasible or dangerous.
8. Select the design that best fits your study. Determine the number of factors to be studied; review the cost and complexity of control of the experiment; consider the need for reflection and for replication.
9. Randomize the order of the DOE as best you can without severe penalty. (And, if you cannot randomize completely, make sure that you are aware of and sensitive to the potential impact of this lack of randomization.) This protects against drawing false conclusions because some factor outside the experiment influenced the response at the same time that a DOE factor level was changed. This is the insurance policy for unbiased conclusions.
10. Perform the experiment. Make sure levels are correctly set for each trial and that materials are identified. Keep good records on all aspects of the DOE.
11. Report the conclusions and recommendations in the language of the audience, not in the terminology of the statistical specialist. Graphs communicate quite well; use them extensively.

10.3 OBSTACLES TO THE APPLICATION OF DOE

Despite its power and versatility, there are several reasons why DOE is not more commonly used. The following is a short list of reasons, not necessarily in order of priority:

1. *Inertia.* "We've always studied one factor at a time and I don't have time to learn a new approach." Remember, if you always do what you've always done, you will always get what you've always gotten!

2. *Technical experts.* This hits every one of us. If we are so confident that we already know all the answers, why is this process unsatisfactory? If we are to achieve breakthrough in any technology (manufacturing, marketing, sales, research, and so on), we must be willing to challenge what we believe. Remember that sometimes we do know the answers, sometimes we believe firmly in myths, and sometimes we don't even know the question! Have you ever noticed that major breakthroughs are often ranked low on the initial Pareto chart of potential answers?

3. *Cost.* How often do we spend a fortune on rework, scrap, or poor sales results but can't afford a DOE to factually resolve the problem once and for all? Nothing is free. DOE is an investment with a payback to be measured just like a capital project.

4. *Manpower.* The major cost of a DOE is often not in materials or lost process time, but in people's time. This is a resource to be managed and prioritized against a return similarly to a capital justification.

5. *Lack of management support.* DOE is a management tool; it cannot be used by line personnel without the concurrence and support of management. Since DOE requires commitment of resources and discipline in performance of the trials, management must be involved. That means management must be educated to the point of understanding what the objectives are, what the costs will be, what should be expected from the study, and what the potential power of DOE is.

6. *Training.* Without a few people educated in the techniques, it is difficult, if not impossible, to have lasting and ongoing impact from DOE.

7. *The assumption that DOE is only a manufacturing tool.* DOE is very useful in determining what actions or strategies are effective in increasing sales or market share, or answering any other question where options can be defined as factors for study. Is it more difficult in nonmanufacturing areas? Probably. But the payoff is also greater!

10.4 DOE SPREADSHEETS AND SOFTWARE CONSIDERATIONS

Now that the basics have been covered—with only the use of an inexpensive scientific calculator—it is time to discuss the use of spreadsheets and software programs. With an understanding of the basic techniques, there are several software packages that are helpful in carrying out the detailed calculations. This author strongly advises using spreadsheets or a dedicated software program to perform and analyze DOE. Doing so will help prevent simple calculation errors and generally speed up the process.

All of the calculations performed in this book could be generated very quickly in a spreadsheet. Even the charts can be made with relative ease. Such spreadsheets can be customized to the experimenter's liking. There are several "free" Microsoft Excel DOE spreadsheets that can be downloaded from the Internet with a simple search. I strongly advise that you verify the spreadsheet to ensure its accuracy so as not to make the wrong decision regarding the process parameters.

There are several commercial DOE software products available. I suggest downloading the "trial" version offered by most software providers and testing the software program prior to purchasing to ensure that it will fulfill the needs of your company.

11
Conclusion

This has been an introduction to design of experiments, the purpose of which was to make the reader capable of handling a majority of the DOE requirements that a technician or engineer might come across. I strongly suggest working the examples provided in this text as a learning exercise prior to conducting an actual experiment. There is much more material to be studied for those interested in more depth. Such things as optimization designs and evolutionary operation (EVOP) have their place. Several complex issues have been avoided in the interest of simplicity, with little consequence to the validity of the techniques that have been presented. However, if one starts experimenting using what has been covered in this book, the technician or engineer will be able to learn and readily transition to more-complex techniques.

All of the calculations performed in this book were generated with an inexpensive scientific calculator. However, these calculations can be generated very quickly in a spreadsheet or a dedicated software program. Spreadsheets and software programs also offer many different options for calculating, producing charts, and graphing, making the analysis of the experimental data results much easier.

One very inexpensive and fun way to teach the basic concepts of DOE to an improvement team is to perform an experiment at a bowling alley! A simple 2^3 experiment can be done with three factors:

A—Hand (left-handed and right-handed)

B—Ball speed (slow and fast)

C—Ball weight (light and heavy)

Another method is to use an inexpensive XPULT Experimental Catapult, available from Peloton Systems LLC (http://xpult.com/).* This is a great training tool for teaching the principles of experimentation, and easily fits into a briefcase.

There is an overwhelming amount of reference material available. Some of the references in the Bibliography will provide greater depth and clarity in more-complex

* The author is not affiliated with XPULT Experimental Catapult or Peloton Systems LLC.

experimentation. Hopefully, the reader has been persuaded to proceed further into this fascinating field.

Finally, go perform some experiments. After performing some studies, it will be obvious that establishing and maintaining a solid program is neither as difficult nor as daunting as one might assume. Good luck!

Appendix A
Critical Values of the Dean and Dixon Outlier Test

	$\alpha = 0.001$	$\alpha = 0.002$	$\alpha = 0.005$	$\alpha = 0.01$	$\alpha = 0.02$	$\alpha = 0.05$	$\alpha = 0.1$	$\alpha = 0.2$
n	r_{10}							
3	0.999	0.998	0.994	0.988	0.976	0.941	0.886	0.782
4	0.964	0.949	0.921	0.889	0.847	0.766	0.679	0.561
5	0.895	0.869	0.824	0.782	0.729	0.643	0.559	0.452
6	0.822	0.792	0.744	0.698	0.646	0.563	0.484	0.387
7	0.763	0.731	0.681	0.636	0.587	0.507	0.433	0.344

Appendix B
Percentages of the *F*-Distribution

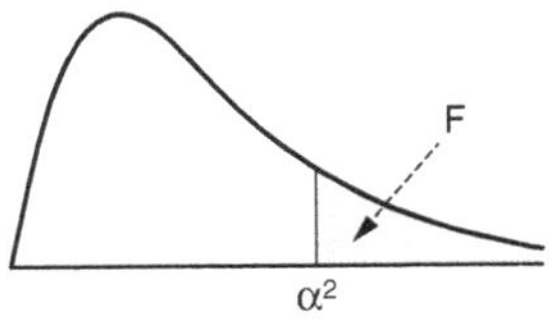

Percentages of the *F*-Distribution α = 0.01										
	Degrees of freedom for numerator ν1									
ν2	1	2	3	4	5	6	7	8	9	10
1	4052	4999	5404	5624	5764	5859	5928	5981	6022	6056
2	98.50	99.00	99.20	99.30	99.30	99.30	99.40	99.40	99.40	99.40
3	34.10	30.80	29.50	28.70	28.20	27.90	27.70	27.50	27.30	27.20
4	21.20	18.00	16.70	16.00	15.50	15.20	15.00	14.80	14.70	14.50
5	16.30	13.30	12.10	11.40	11.00	10.70	10.50	10.30	10.20	10.10
6	13.70	10.90	9.78	9.15	8.75	8.47	8.26	8.10	7.98	7.87
7	12.20	9.55	8.45	7.85	7.46	7.19	6.99	6.84	6.72	6.62
8	11.30	8.65	7.59	7.01	6.63	6.37	6.18	6.03	5.91	5.81
9	10.60	8.02	6.99	6.42	6.06	5.80	5.61	5.47	5.35	5.26
10	10.00	7.56	6.55	5.99	5.64	5.39	5.20	5.06	4.94	4.85
11	9.65	7.21	6.22	5.67	5.32	5.07	4.89	4.74	4.63	4.54
12	9.33	6.93	5.95	5.41	5.06	4.82	4.64	4.50	4.39	4.30
13	9.07	6.70	5.74	5.21	4.86	4.62	4.44	4.30	4.19	4.10
14	8.86	6.51	5.56	5.04	4.69	4.46	4.28	4.14	4.03	3.94
15	8.68	6.36	5.42	4.89	4.56	4.32	4.14	4.00	3.89	3.80
16	8.53	6.23	5.29	4.77	4.44	4.20	4.03	3.89	3.78	3.69
17	8.40	6.11	5.19	4.67	4.34	4.10	3.93	3.79	3.68	3.59
18	8.29	6.01	5.09	4.58	4.25	4.01	3.84	3.71	3.60	3.51
19	8.18	5.93	5.01	4.50	4.17	3.94	3.77	3.63	3.52	3.43
20	8.10	5.85	4.94	4.43	4.10	3.87	3.70	3.56	3.46	3.37
21	8.02	5.78	4.87	4.37	4.04	3.81	3.64	3.51	3.40	3.31
22	7.95	5.72	4.82	4.31	3.99	3.76	3.59	3.45	3.35	3.26
23	7.88	5.66	4.76	4.26	3.94	3.71	3.54	3.41	3.30	3.21
24	7.82	5.61	4.72	4.22	3.90	3.67	3.50	3.36	3.26	3.17
25	7.77	5.57	4.68	4.18	3.85	3.63	3.46	3.32	3.22	3.13
26	7.72	5.53	4.64	4.14	3.82	3.59	3.42	3.29	3.18	3.09
27	7.68	5.49	4.60	4.11	3.78	3.56	3.39	3.26	3.15	3.06
28	7.64	5.45	4.57	4.07	3.75	3.53	3.36	3.23	3.12	3.03
29	7.60	5.42	4.54	4.04	3.73	3.50	3.33	3.20	3.09	3.00
30	7.56	5.39	4.51	4.02	3.70	3.47	3.30	3.17	3.07	2.98
40	7.31	5.18	4.31	3.83	3.51	3.29	3.12	2.99	2.89	2.80
60	7.08	4.98	4.13	3.65	3.34	3.12	2.95	2.82	2.72	2.63
120	6.85	4.79	3.95	3.48	3.17	2.96	2.79	2.66	2.56	2.47
∞	6.64	4.61	3.78	3.32	3.02	2.80	2.64	2.51	2.41	2.32

Continued

Percentages of the *F*-Distribution $\alpha = 0.01$

	Degrees of freedom for numerator ν1								
ν2	**12**	**15**	**20**	**24**	**30**	**40**	**60**	**120**	**∞**
1	6107	6157	6209	6234	6260	6286	6313	6340	6366
2	99.40	99.40	99.40	99.50	99.50	99.50	99.50	99.50	99.50
3	27.10	26.90	26.70	26.60	26.50	26.40	26.30	26.20	26.10
4	14.40	14.20	14.00	13.90	13.80	13.70	13.70	13.60	13.50
5	9.89	9.72	9.55	9.47	9.38	9.29	9.20	9.11	9.02
6	7.72	7.56	7.40	7.31	7.23	7.14	7.06	6.97	6.88
7	6.47	6.31	6.16	6.07	5.99	5.91	5.82	5.74	5.65
8	5.67	5.52	5.36	5.28	5.20	5.12	5.03	4.95	4.86
9	5.11	4.96	4.81	4.73	4.65	4.57	4.48	4.40	4.31
10	4.71	4.56	4.41	4.33	4.25	4.17	4.08	4.00	3.91
11	4.40	4.25	4.10	4.02	3.94	3.86	3.78	3.69	3.60
12	4.16	4.01	3.86	3.78	3.70	3.62	3.54	3.45	3.36
13	3.96	3.82	3.66	3.59	3.51	3.43	3.34	3.25	3.17
14	3.80	3.66	3.51	3.43	3.35	3.27	3.18	3.09	3.00
15	3.67	3.52	3.37	3.29	3.21	3.13	3.05	2.96	2.87
16	3.55	3.41	3.26	3.18	3.10	3.02	2.93	2.84	2.75
17	3.46	3.31	3.16	3.08	3.00	2.92	2.83	2.75	2.65
18	3.37	3.23	3.08	3.00	2.92	2.84	2.75	2.66	2.57
19	3.30	3.15	3.00	2.92	2.84	2.76	2.67	2.58	2.49
20	3.23	3.09	2.94	2.86	2.78	2.69	2.61	2.52	2.42
21	3.17	3.03	2.88	2.80	2.72	2.64	2.55	2.46	2.36
22	3.12	2.98	2.83	2.75	2.67	2.58	2.50	2.40	2.31
23	3.07	2.93	2.78	2.70	2.62	2.54	2.45	2.35	2.26
24	3.03	2.89	2.74	2.66	2.58	2.49	2.40	2.31	2.21
25	2.99	2.85	2.70	2.62	2.54	2.45	2.36	2.27	2.17
26	2.96	2.81	2.66	2.58	2.50	2.42	2.33	2.23	2.13
27	2.93	2.78	2.63	2.55	2.47	2.38	2.29	2.20	2.10
28	2.90	2.75	2.60	2.52	2.44	2.35	2.26	2.17	2.06
29	2.87	2.73	2.57	2.49	2.41	2.33	2.23	2.14	2.03
30	2.84	2.70	2.55	2.47	2.39	2.30	2.21	2.11	2.01
40	2.66	2.52	2.37	2.29	2.20	2.11	2.02	1.92	1.80
60	2.50	2.35	2.20	2.12	2.03	1.94	1.84	1.73	1.60
120	2.34	2.19	2.03	1.95	1.86	1.76	1.66	1.53	1.38
∞	2.18	2.04	1.88	1.79	1.70	1.59	1.47	1.32	1.00

Percentages of the *F*-Distribution α = 0.025										
	Degrees of freedom for numerator ν1									
ν2	1	2	3	4	5	6	7	8	9	10
1	648	799	864	900	922	937	948	957	963	969
2	38.50	39.00	39.20	39.20	39.30	39.30	39.40	39.40	39.40	39.40
3	17.40	16.00	15.40	15.10	14.90	14.70	14.60	14.50	14.50	14.40
4	12.20	10.60	9.98	9.60	9.36	9.20	9.07	8.98	8.90	8.84
5	10.00	8.43	7.76	7.39	7.15	6.98	6.85	6.76	6.68	6.62
6	8.81	7.26	6.60	6.23	5.99	5.82	5.70	5.60	5.52	5.46
7	8.07	6.54	5.89	5.52	5.29	5.12	4.99	4.90	4.82	4.76
8	7.57	6.06	5.42	5.05	4.82	4.65	4.53	4.43	4.36	4.30
9	7.21	5.71	5.08	4.72	4.48	4.32	4.20	4.10	4.03	3.96
10	6.94	5.46	4.83	4.47	4.24	4.07	3.95	3.85	3.78	3.72
11	6.72	5.26	4.63	4.28	4.04	3.88	3.76	3.66	3.59	3.53
12	6.55	5.10	4.47	4.12	3.89	3.73	3.61	3.51	3.44	3.37
13	6.41	4.97	4.35	4.00	3.77	3.60	3.48	3.39	3.31	3.25
14	6.30	4.86	4.24	3.89	3.66	3.50	3.38	3.29	3.21	3.15
15	6.20	4.77	4.15	3.80	3.58	3.41	3.29	3.20	3.12	3.06
16	6.12	4.69	4.08	3.73	3.50	3.34	3.22	3.12	3.05	2.99
17	6.04	4.62	4.01	3.66	3.44	3.28	3.16	3.06	2.98	2.92
18	5.98	4.56	3.95	3.61	3.38	3.22	3.10	3.01	2.93	2.87
19	5.92	4.51	3.90	3.56	3.33	3.17	3.05	2.96	2.88	2.82
20	5.87	4.46	3.86	3.51	3.29	3.13	3.01	2.91	2.84	2.77
21	5.83	4.42	3.82	3.48	3.25	3.09	2.97	2.87	2.80	2.73
22	5.79	4.38	3.78	3.44	3.22	3.05	2.93	2.84	2.76	2.70
23	5.75	4.35	3.75	3.41	3.18	3.02	2.90	2.81	2.73	2.67
24	5.72	4.32	3.72	3.38	3.15	2.99	2.87	2.78	2.70	2.64
25	5.69	4.29	3.69	3.35	3.13	2.97	2.85	2.75	2.68	2.61
26	5.66	4.27	3.67	3.33	3.10	2.94	2.82	2.73	2.65	2.59
27	5.63	4.24	3.65	3.31	3.08	2.92	2.80	2.71	2.63	2.57
28	5.61	4.22	3.63	3.29	3.06	2.90	2.78	2.69	2.61	2.55
29	5.59	4.20	3.61	3.27	3.04	2.88	2.76	2.67	2.59	2.53
30	5.57	4.18	3.59	3.25	3.03	2.87	2.75	2.65	2.57	2.51
40	5.42	4.05	3.46	3.13	2.90	2.74	2.62	2.53	2.45	2.39
60	5.29	3.93	3.34	3.01	2.79	2.63	2.51	2.41	2.33	2.27
120	5.15	3.80	3.23	2.89	2.67	2.52	2.39	2.30	2.22	2.16
∞	5.02	3.69	3.12	2.79	2.57	2.41	2.29	2.19	2.11	2.05

Continued

Percentages of the *F*-Distribution α = 0.025

	Degrees of freedom for numerator ν1								
ν2	12	15	20	24	30	40	60	120	∞
1	977	985	993	997	1001	1006	1010	1014	1018
2	39.40	39.40	39.40	39.50	39.50	39.50	39.50	39.50	39.50
3	14.30	14.30	14.20	14.10	14.10	14.00	14.00	13.90	13.90
4	8.75	8.66	8.56	8.51	8.46	8.41	8.36	8.31	8.26
5	6.52	6.43	6.33	6.28	6.23	6.18	6.12	6.07	6.02
6	5.37	5.27	5.17	5.12	5.07	5.01	4.96	4.90	4.85
7	4.67	4.57	4.47	4.41	4.36	4.31	4.25	4.20	4.14
8	4.20	4.10	4.00	3.95	3.89	3.84	3.78	3.73	3.67
9	3.87	3.77	3.67	3.61	3.56	3.51	3.45	3.39	3.33
10	3.62	3.52	3.42	3.37	3.31	3.26	3.20	3.14	3.08
11	3.43	3.33	3.23	3.17	3.12	3.06	3.00	2.94	2.88
12	3.28	3.18	3.07	3.02	2.96	2.91	2.85	2.79	2.73
13	3.15	3.05	2.95	2.89	2.84	2.78	2.72	2.66	2.60
14	3.05	2.95	2.84	2.79	2.73	2.67	2.61	2.55	2.49
15	2.96	2.86	2.76	2.70	2.64	2.59	2.52	2.46	2.40
16	2.89	2.79	2.68	2.63	2.57	2.51	2.45	2.38	2.32
17	2.82	2.72	2.62	2.56	2.50	2.44	2.38	2.32	2.25
18	2.77	2.67	2.56	2.50	2.44	2.38	2.32	2.26	2.19
19	2.72	2.62	2.51	2.45	2.39	2.33	2.27	2.20	2.13
20	2.68	2.57	2.46	2.41	2.35	2.29	2.22	2.16	2.09
21	2.64	2.53	2.42	2.37	2.31	2.25	2.18	2.11	2.04
22	2.60	2.50	2.39	2.33	2.27	2.21	2.14	2.08	2.00
23	2.57	2.47	2.36	2.30	2.24	2.18	2.11	2.04	1.97
24	2.54	2.44	2.33	2.27	2.21	2.15	2.08	2.01	1.94
25	2.51	2.41	2.30	2.24	2.18	2.12	2.05	1.98	1.91
26	2.49	2.39	2.28	2.22	2.16	2.09	2.03	1.95	1.88
27	2.47	2.36	2.25	2.19	2.13	2.07	2.00	1.93	1.85
28	2.45	2.34	2.23	2.17	2.11	2.05	1.98	1.91	1.83
29	2.43	2.32	2.21	2.15	2.09	2.03	1.96	1.89	1.81
30	2.41	2.31	2.20	2.14	2.07	2.01	1.94	1.87	1.79
40	2.29	2.18	2.07	2.01	1.94	1.88	1.80	1.72	1.64
60	2.17	2.06	1.94	1.88	1.82	1.74	1.67	1.58	1.48
120	2.05	1.94	1.82	1.76	1.69	1.61	1.53	1.43	1.31
∞	1.94	1.83	1.71	1.64	1.57	1.48	1.39	1.27	1.00

Percentages of the *F*-Distribution $\alpha = 0.05$										
	Degrees of freedom for numerator $\nu 1$									
$\nu 2$	1	2	3	4	5	6	7	8	9	10
1	161	199	216	225	230	234	237	239	241	242
2	18.50	19.00	19.20	19.20	19.30	19.30	19.40	19.40	19.40	19.40
3	10.10	9.55	9.28	9.12	9.01	8.94	8.89	8.85	8.81	8.79
4	7.71	6.94	6.59	6.39	6.26	6.16	6.09	6.04	6.00	5.96
5	6.61	5.79	5.41	5.19	5.05	4.95	4.88	4.82	4.77	4.74
6	5.99	5.14	4.76	4.53	4.39	4.28	4.21	4.15	4.10	4.06
7	5.59	4.74	4.35	4.12	3.97	3.87	3.79	3.73	3.68	3.64
8	5.32	4.46	4.07	3.84	3.69	3.58	3.50	3.44	3.39	3.35
9	5.12	4.26	3.86	3.63	3.48	3.37	3.29	3.23	3.18	3.14
10	4.96	4.10	3.71	3.48	3.33	3.22	3.14	3.07	3.02	2.98
11	4.84	3.98	3.59	3.36	3.20	3.09	3.01	2.95	2.90	2.85
12	4.75	3.89	3.49	3.26	3.11	3.00	2.91	2.85	2.80	2.75
13	4.67	3.81	3.41	3.18	3.03	2.92	2.83	2.77	2.71	2.67
14	4.60	3.74	3.34	3.11	2.96	2.85	2.76	2.70	2.65	2.60
15	4.54	3.68	3.29	3.06	2.90	2.79	2.71	2.64	2.59	2.54
16	4.49	3.63	3.24	3.01	2.85	2.74	2.66	2.59	2.54	2.49
17	4.45	3.59	3.20	2.96	2.81	2.70	2.61	2.55	2.49	2.45
18	4.41	3.55	3.16	2.93	2.77	2.66	2.58	2.51	2.46	2.41
19	4.38	3.52	3.13	2.90	2.74	2.63	2.54	2.48	2.42	2.38
20	4.35	3.49	3.10	2.87	2.71	2.60	2.51	2.45	2.39	2.35
21	4.32	3.47	3.07	2.84	2.68	2.57	2.49	2.42	2.37	2.32
22	4.30	3.44	3.05	2.82	2.66	2.55	2.46	2.40	2.34	2.30
23	4.28	3.42	3.03	2.80	2.64	2.53	2.44	2.37	2.32	2.27
24	4.26	3.40	3.01	2.78	2.62	2.51	2.42	2.36	2.30	2.25
25	4.24	3.39	2.99	2.76	2.60	2.49	2.40	2.34	2.28	2.24
26	4.23	3.37	2.98	2.74	2.59	2.47	2.39	2.32	2.27	2.22
27	4.21	3.35	2.96	2.73	2.57	2.46	2.37	2.31	2.25	2.20
28	4.20	3.34	2.95	2.71	2.56	2.45	2.36	2.29	2.24	2.19
29	4.18	3.33	2.93	2.70	2.55	2.43	2.35	2.28	2.22	2.18
30	4.17	3.32	2.92	2.69	2.53	2.42	2.33	2.27	2.21	2.16
40	4.08	3.23	2.84	2.61	2.45	2.34	2.25	2.18	2.12	2.08
60	4.00	3.15	2.76	2.53	2.37	2.25	2.17	2.10	2.04	1.99
120	3.92	3.07	2.68	2.45	2.29	2.18	2.09	2.02	1.96	1.91
∞	3.84	3.00	2.60	2.37	2.21	2.10	2.01	1.94	1.88	1.83

Continued

Percentages of the *F*-Distribution α = 0.05

	Degrees of freedom for numerator ν1								
ν2	**12**	**15**	**20**	**24**	**30**	**40**	**60**	**120**	**∞**
1	244	246	248	249	250	251	252	253	254
2	19.40	19.40	19.40	19.50	19.50	19.50	19.50	19.50	19.50
3	8.74	8.70	8.66	8.64	8.62	8.59	8.57	8.55	8.53
4	5.91	5.86	5.80	5.77	5.75	5.72	5.69	5.66	5.63
5	4.68	4.62	4.56	4.53	4.50	4.46	4.43	4.40	4.37
6	4.00	3.94	3.87	3.84	3.81	3.77	3.74	3.70	3.67
7	3.57	3.51	3.44	3.41	3.38	3.34	3.30	3.27	3.23
8	3.28	3.22	3.15	3.12	3.08	3.04	3.01	2.97	2.93
9	3.07	3.01	2.94	2.90	2.86	2.83	2.79	2.75	2.71
10	2.91	2.85	2.77	2.74	2.70	2.66	2.62	2.58	2.54
11	2.79	2.72	2.65	2.61	2.57	2.53	2.49	2.45	2.40
12	2.69	2.62	2.54	2.51	2.47	2.43	2.38	2.34	2.30
13	2.60	2.53	2.46	2.42	2.38	2.34	2.30	2.25	2.21
14	2.53	2.46	2.39	2.35	2.31	2.27	2.22	2.18	2.13
15	2.48	2.40	2.33	2.29	2.25	2.20	2.16	2.11	2.07
16	2.42	2.35	2.28	2.24	2.19	2.15	2.11	2.06	2.01
17	2.38	2.31	2.23	2.19	2.15	2.10	2.06	2.01	1.96
18	2.34	2.27	2.19	2.15	2.11	2.06	2.02	1.97	1.92
19	2.31	2.23	2.16	2.11	2.07	2.03	1.98	1.93	1.88
20	2.28	2.20	2.12	2.08	2.04	1.99	1.95	1.90	1.84
21	2.25	2.18	2.10	2.05	2.01	1.96	1.92	1.87	1.81
22	2.23	2.15	2.07	2.03	1.98	1.94	1.89	1.84	1.78
23	2.20	2.13	2.05	2.01	1.96	1.91	1.86	1.81	1.76
24	2.18	2.11	2.03	1.98	1.94	1.89	1.84	1.79	1.73
25	2.16	2.09	2.01	1.96	1.92	1.87	1.82	1.77	1.71
26	2.15	2.07	1.99	1.95	1.90	1.85	1.80	1.75	1.69
27	2.13	2.06	1.97	1.93	1.88	1.84	1.79	1.73	1.67
28	2.12	2.04	1.96	1.91	1.87	1.82	1.77	1.71	1.65
29	2.10	2.03	1.94	1.90	1.85	1.81	1.75	1.70	1.64
30	2.09	2.01	1.93	1.89	1.84	1.79	1.74	1.68	1.62
40	2.00	1.92	1.84	1.79	1.74	1.69	1.64	1.58	1.51
60	1.92	1.84	1.75	1.70	1.65	1.59	1.53	1.47	1.39
120	1.83	1.75	1.66	1.61	1.55	1.50	1.43	1.35	1.25
∞	1.75	1.67	1.57	1.52	1.46	1.39	1.32	1.22	1.00

Percentages of the *F*-Distribution α = 0.10										
	Degrees of freedom for numerator ν1									
ν2	1	2	3	4	5	6	7	8	9	10
1	40	50	54	56	57	58	59	59	60	60
2	8.53	9.00	9.16	9.24	9.29	9.33	9.35	9.37	9.38	9.39
3	5.54	5.46	5.39	5.34	5.31	5.28	5.27	5.25	5.24	5.23
4	4.54	4.32	4.19	4.11	4.05	4.01	3.98	3.95	3.94	3.92
5	4.06	3.78	3.62	3.52	3.45	3.40	3.37	3.34	3.32	3.30
6	3.78	3.46	3.29	3.18	3.11	3.05	3.01	2.98	2.96	2.94
7	3.59	3.26	3.07	2.96	2.88	2.83	2.78	2.75	2.72	2.70
8	3.46	3.11	2.92	2.81	2.73	2.67	2.62	2.59	2.56	2.54
9	3.36	3.01	2.81	2.69	2.61	2.55	2.51	2.47	2.44	2.42
10	3.29	2.92	2.73	2.61	2.52	2.46	2.41	2.38	2.35	2.32
11	3.23	2.86	2.66	2.54	2.45	2.39	2.34	2.30	2.27	2.25
12	3.18	2.81	2.61	2.48	2.39	2.33	2.28	2.24	2.21	2.19
13	3.14	2.76	2.56	2.43	2.35	2.28	2.23	2.20	2.16	2.14
14	3.10	2.73	2.52	2.39	2.31	2.24	2.19	2.15	2.12	2.10
15	3.07	2.70	2.49	2.36	2.27	2.21	2.16	2.12	2.09	2.06
16	3.05	2.67	2.46	2.33	2.24	2.18	2.13	2.09	2.06	2.03
17	3.03	2.64	2.44	2.31	2.22	2.15	2.10	2.06	2.03	2.00
18	3.01	2.62	2.42	2.29	2.20	2.13	2.08	2.04	2.00	1.98
19	2.99	2.61	2.40	2.27	2.18	2.11	2.06	2.02	1.98	1.96
20	2.97	2.59	2.38	2.25	2.16	2.09	2.04	2.00	1.96	1.94
21	2.96	2.57	2.36	2.23	2.14	2.08	2.02	1.98	1.95	1.92
22	2.95	2.56	2.35	2.22	2.13	2.06	2.01	1.97	1.93	1.90
23	2.94	2.55	2.34	2.21	2.11	2.05	1.99	1.95	1.92	1.89
24	2.93	2.54	2.33	2.19	2.10	2.04	1.98	1.94	1.91	1.88
25	2.92	2.53	2.32	2.18	2.09	2.02	1.97	1.93	1.89	1.87
26	2.91	2.52	2.31	2.17	2.08	2.01	1.96	1.92	1.88	1.86
27	2.90	2.51	2.30	2.17	2.07	2.00	1.95	1.91	1.87	1.85
28	2.89	2.50	2.29	2.16	2.06	2.00	1.94	1.90	1.87	1.84
29	2.89	2.50	2.28	2.15	2.06	1.99	1.93	1.89	1.86	1.83
30	2.88	2.49	2.28	2.14	2.05	1.98	1.93	1.88	1.85	1.82
40	2.84	2.44	2.23	2.09	2.00	1.93	1.87	1.83	1.79	1.76
60	2.79	2.39	2.18	2.04	1.95	1.87	1.82	1.77	1.74	1.71
120	2.75	2.35	2.13	1.99	1.90	1.82	1.77	1.72	1.68	1.65
∞	2.71	2.30	2.08	1.94	1.85	1.77	1.72	1.67	1.63	1.60

Continued

Percentages of the *F*-Distribution α = 0.10									
	Degrees of freedom for numerator ν1								
ν2	12	15	20	24	30	40	60	120	∞
1	61	61	62	62	62	63	63	63	63
2	9.41	9.42	9.44	9.45	9.46	9.47	9.47	9.48	9.49
3	5.22	5.20	5.18	5.18	5.17	5.16	5.15	5.14	5.13
4	3.90	3.87	3.84	3.83	3.82	3.80	3.79	3.78	3.76
5	3.27	3.24	3.21	3.19	3.17	3.16	3.14	3.12	3.11
6	2.90	2.87	2.84	2.82	2.80	2.78	2.76	2.74	2.72
7	2.67	2.63	2.59	2.58	2.56	2.54	2.51	2.49	2.47
8	2.50	2.46	2.42	2.40	2.38	2.36	2.34	2.32	2.29
9	2.38	2.34	2.30	2.28	2.25	2.23	2.21	2.18	2.16
10	2.28	2.24	2.20	2.18	2.16	2.13	2.11	2.08	2.06
11	2.21	2.17	2.12	2.10	2.08	2.05	2.03	2.00	1.97
12	2.15	2.10	2.06	2.04	2.01	1.99	1.96	1.93	1.90
13	2.10	2.05	2.01	1.98	1.96	1.93	1.90	1.88	1.85
14	2.05	2.01	1.96	1.94	1.91	1.89	1.86	1.83	1.80
15	2.02	1.97	1.92	1.90	1.87	1.85	1.82	1.79	1.76
16	1.99	1.94	1.89	1.87	1.84	1.81	1.78	1.75	1.72
17	1.96	1.91	1.86	1.84	1.81	1.78	1.75	1.72	1.69
18	1.93	1.89	1.84	1.81	1.78	1.75	1.72	1.69	1.66
19	1.91	1.86	1.81	1.79	1.76	1.73	1.70	1.67	1.63
20	1.89	1.84	1.79	1.77	1.74	1.71	1.68	1.64	1.61
21	1.87	1.83	1.78	1.75	1.72	1.69	1.66	1.62	1.59
22	1.86	1.81	1.76	1.73	1.70	1.67	1.64	1.60	1.57
23	1.84	1.80	1.74	1.72	1.69	1.66	1.62	1.59	1.55
24	1.83	1.78	1.73	1.70	1.67	1.64	1.61	1.57	1.53
25	1.82	1.77	1.72	1.69	1.66	1.63	1.59	1.56	1.52
26	1.81	1.76	1.71	1.68	1.65	1.61	1.58	1.54	1.50
27	1.80	1.75	1.70	1.67	1.64	1.60	1.57	1.53	1.49
28	1.79	1.74	1.69	1.66	1.63	1.59	1.56	1.52	1.48
29	1.78	1.73	1.68	1.65	1.62	1.58	1.55	1.51	1.47
30	1.77	1.72	1.67	1.64	1.61	1.57	1.54	1.50	1.46
40	1.71	1.66	1.61	1.57	1.54	1.51	1.47	1.42	1.38
60	1.66	1.60	1.54	1.51	1.48	1.44	1.40	1.35	1.29
120	1.60	1.55	1.48	1.45	1.41	1.37	1.32	1.26	1.19
∞	1.55	1.49	1.42	1.38	1.34	1.30	1.24	1.17	1.00

Percentages of the *F*-Distribution α = 0.25										
	Degrees of freedom for numerator ν1									
ν2	1	2	3	4	5	6	7	8	9	10
1	6	8	8	9	9	9	9	9	9	9
2	2.57	3.00	3.15	3.23	3.28	3.31	3.34	3.35	3.37	3.38
3	2.02	2.28	2.36	2.39	2.41	2.42	2.43	2.44	2.44	2.44
4	1.81	2.00	2.05	2.06	2.07	2.08	2.08	2.08	2.08	2.08
5	1.69	1.85	1.88	1.89	1.89	1.89	1.89	1.89	1.89	1.89
6	1.62	1.76	1.78	1.79	1.79	1.78	1.78	1.78	1.77	1.77
7	1.57	1.70	1.72	1.72	1.71	1.71	1.70	1.70	1.69	1.69
8	1.54	1.66	1.67	1.66	1.66	1.65	1.64	1.64	1.63	1.63
9	1.51	1.62	1.63	1.63	1.62	1.61	1.60	1.60	1.59	1.59
10	1.49	1.60	1.60	1.59	1.59	1.58	1.57	1.56	1.56	1.55
11	1.47	1.58	1.58	1.57	1.56	1.55	1.54	1.53	1.53	1.52
12	1.46	1.56	1.56	1.55	1.54	1.53	1.52	1.51	1.51	1.50
13	1.45	1.55	1.55	1.53	1.52	1.51	1.50	1.49	1.49	1.48
14	1.44	1.53	1.53	1.52	1.51	1.50	1.49	1.48	1.47	1.46
15	1.43	1.52	1.52	1.51	1.49	1.48	1.47	1.46	1.46	1.45
16	1.42	1.51	1.51	1.50	1.48	1.47	1.46	1.45	1.44	1.44
17	1.42	1.51	1.50	1.49	1.47	1.46	1.45	1.44	1.43	1.43
18	1.41	1.50	1.49	1.48	1.46	1.45	1.44	1.43	1.42	1.42
19	1.41	1.49	1.49	1.47	1.46	1.44	1.43	1.42	1.41	1.41
20	1.40	1.49	1.48	1.47	1.45	1.44	1.43	1.42	1.41	1.40
21	1.40	1.48	1.48	1.46	1.44	1.43	1.42	1.41	1.40	1.39
22	1.40	1.48	1.47	1.45	1.44	1.42	1.41	1.40	1.39	1.39
23	1.39	1.47	1.47	1.45	1.43	1.42	1.41	1.40	1.39	1.38
24	1.39	1.47	1.46	1.44	1.43	1.41	1.40	1.39	1.38	1.38
25	1.39	1.47	1.46	1.44	1.42	1.41	1.40	1.39	1.38	1.37
26	1.38	1.46	1.45	1.44	1.42	1.41	1.39	1.38	1.37	1.37
27	1.38	1.46	1.45	1.43	1.42	1.40	1.39	1.38	1.37	1.36
28	1.38	1.46	1.45	1.43	1.41	1.40	1.39	1.38	1.37	1.36
29	1.38	1.45	1.45	1.43	1.41	1.40	1.38	1.37	1.36	1.35
30	1.38	1.45	1.44	1.42	1.41	1.39	1.38	1.37	1.36	1.35
40	1.36	1.44	1.42	1.40	1.39	1.37	1.36	1.35	1.34	1.33
60	1.35	1.42	1.41	1.38	1.37	1.35	1.33	1.32	1.31	1.30
120	1.34	1.40	1.39	1.37	1.35	1.33	1.31	1.30	1.29	1.28
∞	1.32	1.39	1.37	1.35	1.33	1.31	1.29	1.28	1.27	1.25

Continued

Percentages of the *F*-Distribution α = 0.25									
	Degrees of freedom for numerator ν1								
ν2	12	15	20	24	30	40	60	120	∞
1	9	9	10	10	10	10	10	10	10
2	3.39	3.41	3.43	3.43	3.44	3.45	3.46	3.47	3.48
3	2.45	2.46	2.46	2.46	2.47	2.47	2.47	2.47	2.47
4	2.08	2.08	2.08	2.08	2.08	2.08	2.08	2.08	2.08
5	1.89	1.89	1.88	1.88	1.88	1.88	1.87	1.87	1.87
6	1.77	1.76	1.76	1.75	1.75	1.75	1.74	1.74	1.74
7	1.68	1.68	1.67	1.67	1.66	1.66	1.65	1.65	1.65
8	1.62	1.62	1.61	1.60	1.60	1.59	1.59	1.58	1.58
9	1.58	1.57	1.56	1.56	1.55	1.54	1.54	1.53	1.53
10	1.54	1.53	1.52	1.52	1.51	1.51	1.50	1.49	1.48
11	1.51	1.50	1.49	1.49	1.48	1.47	1.47	1.46	1.45
12	1.49	1.48	1.47	1.46	1.45	1.45	1.44	1.43	1.42
13	1.47	1.46	1.45	1.44	1.43	1.42	1.42	1.41	1.40
14	1.45	1.44	1.43	1.42	1.41	1.41	1.40	1.39	1.38
15	1.44	1.43	1.41	1.41	1.40	1.39	1.38	1.37	1.36
16	1.43	1.41	1.40	1.39	1.38	1.37	1.36	1.35	1.34
17	1.41	1.40	1.39	1.38	1.37	1.36	1.35	1.34	1.33
18	1.40	1.39	1.38	1.37	1.36	1.35	1.34	1.33	1.32
19	1.40	1.38	1.37	1.36	1.35	1.34	1.33	1.32	1.30
20	1.39	1.37	1.36	1.35	1.34	1.33	1.32	1.31	1.29
21	1.38	1.37	1.35	1.34	1.33	1.32	1.31	1.30	1.28
22	1.37	1.36	1.34	1.33	1.32	1.31	1.30	1.29	1.28
23	1.37	1.35	1.34	1.33	1.32	1.31	1.30	1.28	1.27
24	1.36	1.35	1.33	1.32	1.31	1.30	1.29	1.28	1.26
25	1.36	1.34	1.33	1.32	1.31	1.29	1.28	1.27	1.25
26	1.35	1.34	1.32	1.31	1.30	1.29	1.28	1.26	1.25
27	1.35	1.33	1.32	1.31	1.30	1.28	1.27	1.26	1.24
28	1.34	1.33	1.31	1.30	1.29	1.28	1.27	1.25	1.24
29	1.34	1.32	1.31	1.30	1.29	1.27	1.26	1.25	1.23
30	1.34	1.32	1.30	1.29	1.28	1.27	1.26	1.24	1.23
40	1.31	1.30	1.28	1.26	1.25	1.24	1.22	1.21	1.19
60	1.29	1.27	1.25	1.24	1.22	1.21	1.19	1.17	1.15
120	1.26	1.24	1.22	1.21	1.19	1.18	1.16	1.13	1.10
∞	1.24	1.22	1.19	1.18	1.16	1.14	1.12	1.08	1.00

Appendix C

Percentage Points of the Student's *t*-Distribution

	p	0.9	0.95	0.975	0.9875	0.995	0.9975	0.999	0.9995
df	α	**0.1**	**0.05**	**0.025**	**0.0125**	**0.005**	**0.0025**	**0.001**	**0.0005**
1		3.078	6.314	12.706	25.452	63.657	127.321	318.309	636.619
2		1.886	2.920	4.303	6.205	9.925	14.089	22.327	31.599
3		1.638	2.353	3.182	4.177	5.841	7.453	10.215	12.924
4		1.533	2.132	2.776	3.495	4.604	5.598	7.173	8.610
5		1.476	2.015	2.571	3.163	4.032	4.773	5.893	6.869
6		1.440	1.943	2.447	2.969	3.707	4.317	5.208	5.959
7		1.415	1.895	2.365	2.841	3.499	4.029	4.785	5.408
8		1.397	1.860	2.306	2.752	3.355	3.833	4.501	5.041
9		1.383	1.833	2.262	2.685	3.250	3.690	4.297	4.781
10		1.372	1.812	2.228	2.634	3.169	3.581	4.144	4.587
11		1.363	1.796	2.201	2.593	3.106	3.497	4.025	4.437
12		1.356	1.782	2.179	2.560	3.055	3.428	3.930	4.318
13		1.350	1.771	2.160	2.533	3.012	3.372	3.852	4.221
14		1.345	1.761	2.145	2.510	2.977	3.326	3.787	4.140
15		1.341	1.753	2.131	2.490	2.947	3.286	3.733	4.073
16		1.337	1.746	2.120	2.473	2.921	3.252	3.686	4.015
17		1.333	1.740	2.110	2.458	2.898	3.222	3.646	3.965
18		1.330	1.734	2.101	2.445	2.878	3.197	3.610	3.922
19		1.328	1.729	2.093	2.433	2.861	3.174	3.579	3.883
20		1.325	1.725	2.086	2.423	2.845	3.153	3.552	3.850
21		1.323	1.721	2.080	2.414	2.831	3.135	3.527	3.819

Continued

Continued

	p	0.9	0.95	0.975	0.9875	0.995	0.9975	0.999	0.9995
df	α	**0.1**	**0.05**	**0.025**	**0.0125**	**0.005**	**0.0025**	**0.001**	**0.0005**
22		1.321	1.717	2.074	2.405	2.819	3.119	3.505	3.792
23		1.319	1.714	2.069	2.398	2.807	3.104	3.485	3.768
24		1.318	1.711	2.064	2.391	2.797	3.090	3.467	3.745
25		1.316	1.708	2.060	2.385	2.787	3.078	3.450	3.725
26		1.315	1.706	2.056	2.379	2.779	3.067	3.435	3.707
27		1.314	1.703	2.052	2.373	2.771	3.057	3.421	3.690
28		1.313	1.701	2.048	2.368	2.763	3.047	3.408	3.674
29		1.311	1.699	2.045	2.364	2.756	3.038	3.396	3.659
30		1.310	1.697	2.042	2.360	2.750	3.030	3.385	3.646
60		1.296	1.671	2.000	2.299	2.660	2.915	3.232	3.460
90		1.291	1.662	1.987	2.280	2.632	2.878	3.183	3.402
120		1.289	1.658	1.980	2.270	2.617	2.860	3.160	3.373
∞		1.282	1.645	1.960	2.242	2.576	2.807	3.090	3.291

Appendix D
Cumulative Percentage Points

	Number of effects								
Cumulative percentage	20	19	18	17	16	15	14	13	12
	0.025	0.026	0.028	0.029	0.031	0.033	0.036	0.038	0.042
	0.075	0.079	0.083	0.088	0.094	0.100	0.107	0.115	0.125
	0.125	0.132	0.139	0.147	0.156	0.167	0.179	0.192	0.208
	0.175	0.184	0.194	0.206	0.219	0.233	0.250	0.269	0.292
	0.225	0.237	0.250	0.265	0.281	0.300	0.321	0.346	0.375
	0.275	0.289	0.306	0.324	0.344	0.367	0.393	0.423	0.458
	0.325	0.342	0.361	0.382	0.406	0.433	0.464	0.500	0.542
	0.375	0.395	0.417	0.441	0.469	0.500	0.536	0.577	0.625
	0.425	0.447	0.472	0.500	0.531	0.567	0.607	0.654	0.708
	0.475	0.500	0.528	0.559	0.594	0.633	0.679	0.731	0.792
	0.525	0.553	0.583	0.618	0.656	0.700	0.750	0.808	0.875
	0.575	0.605	0.639	0.676	0.719	0.767	0.821	0.885	0.958
	0.625	0.658	0.694	0.735	0.781	0.833	0.893	0.962	
	0.675	0.711	0.750	0.794	0.844	0.900	0.964		
	0.725	0.763	0.806	0.853	0.906	0.967			
	0.775	0.816	0.861	0.912	0.969				
	0.825	0.868	0.917	0.971					
	0.875	0.921	0.972						
	0.925	0.974							
	0.975								

Cumulative percentage	Number of effects								
	11	10	9	8	7	6	5	4	3
	0.045	0.050	0.056	0.063	0.071	0.083	0.100	0.125	0.167
	0.136	0.150	0.167	0.188	0.214	0.250	0.300	0.375	0.500
	0.227	0.250	0.278	0.313	0.357	0.417	0.500	0.625	0.833
	0.318	0.350	0.389	0.438	0.500	0.583	0.700	0.875	
	0.409	0.450	0.500	0.563	0.643	0.750	0.900		
	0.500	0.550	0.611	0.688	0.786	0.917			
	0.591	0.650	0.722	0.813	0.929				
	0.682	0.750	0.833	0.938					
	0.773	0.850	0.944						
	0.864	0.950							

Appendix E

z-Scores of the Cumulative Percentage Points

	Number of effects								
	20	19	18	17	16	15	14	13	12
z-Score	–1.960	–1.938	–1.915	–1.890	–1.863	–1.834	–1.803	–1.769	–1.732
	–1.440	–1.412	–1.383	–1.352	–1.318	–1.282	–1.242	–1.198	–1.150
	–1.150	–1.119	–1.085	–1.049	–1.010	–0.967	–0.921	–0.869	–0.812
	–0.935	–0.899	–0.862	–0.821	–0.776	–0.728	–0.674	–0.615	–0.549
	–0.755	–0.716	–0.674	–0.629	–0.579	–0.524	–0.464	–0.396	–0.319
	–0.598	–0.555	–0.508	–0.458	–0.402	–0.341	–0.272	–0.194	–0.105
	–0.454	–0.407	–0.355	–0.299	–0.237	–0.168	–0.090	0.000	0.105
	–0.319	–0.267	–0.210	–0.148	–0.078	0.000	0.090	0.194	0.319
	–0.189	–0.132	–0.070	0.000	0.078	0.168	0.272	0.396	0.549
	–0.063	0.000	0.070	0.148	0.237	0.341	0.464	0.615	0.812
	0.063	0.132	0.210	0.299	0.402	0.524	0.674	0.869	1.150
	0.189	0.267	0.355	0.458	0.579	0.728	0.921	1.198	1.732
	0.319	0.407	0.508	0.629	0.776	0.967	1.242	1.769	
	0.454	0.555	0.674	0.821	1.010	1.282	1.803		
	0.598	0.716	0.862	1.049	1.318	1.834			
	0.755	0.899	1.085	1.352	1.863				
	0.935	1.119	1.383	1.890					
	1.150	1.412	1.915						
	1.440	1.938							
	1.960								

	Number of effects								
	11	10	9	8	7	6	5	4	3
z-Score	−1.691	−1.645	−1.593	−1.534	−1.465	−1.383	−1.282	−1.150	−0.967
	−1.097	−1.036	−0.967	−0.887	−0.792	−0.674	−0.524	−0.319	0.000
	−0.748	−0.674	−0.589	−0.489	−0.366	−0.210	0.000	0.319	0.967
	−0.473	−0.385	−0.282	−0.157	0.000	0.210	0.524	1.150	
	−0.230	−0.126	0.000	0.157	0.366	0.674	1.282		
	0.000	0.126	0.282	0.489	0.792	1.383			
	0.230	0.385	0.589	0.887	1.465				
	0.473	0.674	0.967	1.534					
	0.748	1.036	1.593						
	1.097	1.645							
	1.691								

Appendix F

Normal Distribution Probability Points—Area below *Z*

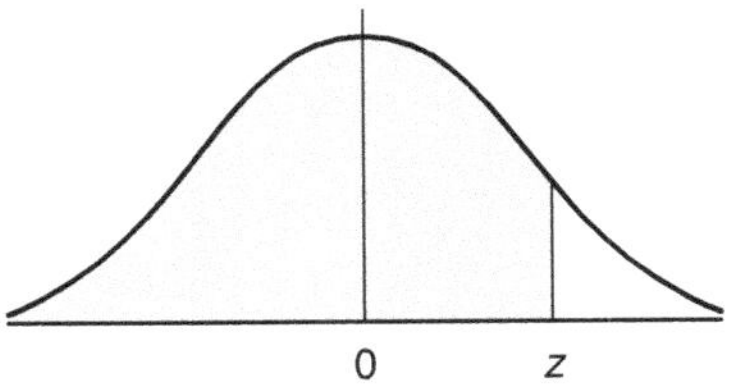

z	0	0.01	0.02	0.03	0.04	0.05	0.06	0.07	0.08	0.09
0	.5000	.5040	.5080	.5120	.5160	.5199	.5239	.5279	.5319	.5359
0.1	.5398	.5438	.5478	.5517	.5557	.5596	.5636	.5675	.5714	.5753
0.2	.5793	.5832	.5871	.5910	.5948	.5987	.6026	.6064	.6103	.6141
0.3	.6179	.6217	.6255	.6293	.6331	.6368	.6406	.6443	.6480	.6517
0.4	.6554	.6591	.6628	.6664	.6700	.6736	.6772	.6808	.6844	.6879
0.5	.6915	.6950	.6985	.7019	.7054	.7088	.7123	.7157	.7190	.7224
0.6	.7257	.7291	.7324	.7357	.7389	.7422	.7454	.7486	.7517	.7549
0.7	.7580	.7611	.7642	.7673	.7704	.7734	.7764	.7794	.7823	.7852
0.8	.7881	.7910	.7939	.7967	.7995	.8023	.8051	.8078	.8106	.8133
0.9	.8159	.8186	.8212	.8238	.8264	.8289	.8315	.8340	.8365	.8389
1	.8413	.8438	.8461	.8485	.8508	.8531	.8554	.8577	.8599	.8621
1.1	.8643	.8665	.8686	.8708	.8729	.8749	.8770	.8790	.8810	.8830
1.2	.8849	.8869	.8888	.8907	.8925	.8944	.8962	.8980	.8997	.9015
1.3	.9032	.9049	.9066	.9082	.9099	.9115	.9131	.9147	.9162	.9177
1.4	.9192	.9207	.9222	.9236	.9251	.9265	.9279	.9292	.9306	.9319
1.5	.9332	.9345	.9357	.9370	.9382	.9394	.9406	.9418	.9429	.9441
1.6	.9452	.9463	.9474	.9484	.9495	.9505	.9515	.9525	.9535	.9545
1.7	.9554	.9564	.9573	.9582	.9591	.9599	.9608	.9616	.9625	.9633
1.8	.9641	.9649	.9656	.9664	.9671	.9678	.9686	.9693	.9699	.9706
1.9	.9713	.9719	.9726	.9732	.9738	.9744	.9750	.9756	.9761	.9767
2	.9772	.9778	.9783	.9788	.9793	.9798	.9803	.9808	.9812	.9817
2.1	.9821	.9826	.9830	.9834	.9838	.9842	.9846	.9850	.9854	.9857
2.2	.9861	.9864	.9868	.9871	.9875	.9878	.9881	.9884	.9887	.9890
2.3	.9893	.9896	.9898	.9901	.9904	.9906	.9909	.9911	.9913	.9916
2.4	.9918	.9920	.9922	.9925	.9927	.9929	.9931	.9932	.9934	.9936
2.5	.9938	.9940	.9941	.9943	.9945	.9946	.9948	.9949	.9951	.9952
2.6	.9953	.9955	.9956	.9957	.9959	.9960	.9961	.9962	.9963	.9964
2.7	.9965	.9966	.9967	.9968	.9969	.9970	.9971	.9972	.9973	.9974
2.8	.9974	.9975	.9976	.9977	.9977	.9978	.9979	.9979	.9980	.9981
2.9	.9981	.9982	.9982	.9983	.9984	.9984	.9985	.9985	.9986	.9986
3	.9987	.9987	.9987	.9988	.9988	.9989	.9989	.9989	.9990	.9990

Appendix G

Normal Distribution Probability Points—Area above *Z*

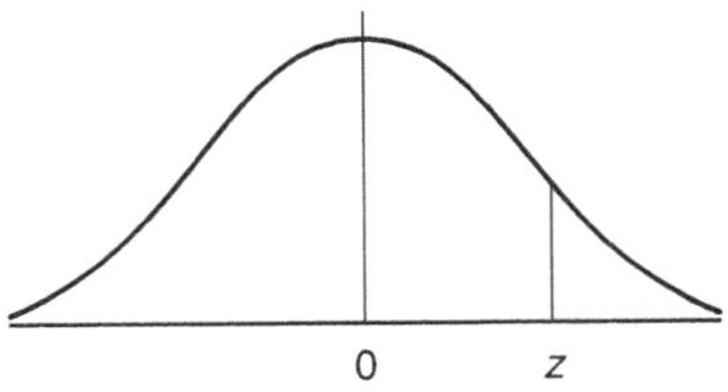

z	0	0.01	0.02	0.03	0.04	0.05	0.06	0.07	0.08	0.09
0	.5000	.4960	.4920	.4880	.4840	.4801	.4761	.4721	.4681	.4641
0.1	.4602	.4562	.4522	.4483	.4443	.4404	.4364	.4325	.4286	.4247
0.2	.4207	.4168	.4129	.4090	.4052	.4013	.3974	.3936	.3897	.3859
0.3	.3821	.3783	.3745	.3707	.3669	.3632	.3594	.3557	.3520	.3483
0.4	.3446	.3409	.3372	.3336	.3300	.3264	.3228	.3192	.3156	.3121
0.5	.3085	.3050	.3015	.2981	.2946	.2912	.2877	.2843	.2810	.2776
0.6	.2743	.2709	.2676	.2643	.2611	.2578	.2546	.2514	.2483	.2451
0.7	.2420	.2389	.2358	.2327	.2296	.2266	.2236	.2206	.2177	.2148
0.8	.2119	.2090	.2061	.2033	.2005	.1977	.1949	.1922	.1894	.1867
0.9	.1841	.1814	.1788	.1762	.1736	.1711	.1685	.1660	.1635	.1611
1	.1587	.1562	.1539	.1515	.1492	.1469	.1446	.1423	.1401	.1379
1.1	.1357	.1335	.1314	.1292	.1271	.1251	.1230	.1210	.1190	.1170
1.2	.1151	.1131	.1112	.1093	.1075	.1056	.1038	.1020	.1003	.0985
1.3	.0968	.0951	.0934	.0918	.0901	.0885	.0869	.0853	.0838	.0823
1.4	.0808	.0793	.0778	.0764	.0749	.0735	.0721	.0708	.0694	.0681
1.5	.0668	.0655	.0643	.0630	.0618	.0606	.0594	.0582	.0571	.0559
1.6	.0548	.0537	.0526	.0516	.0505	.0495	.0485	.0475	.0465	.0455
1.7	.0446	.0436	.0427	.0418	.0409	.0401	.0392	.0384	.0375	.0367
1.8	.0359	.0351	.0344	.0336	.0329	.0322	.0314	.0307	.0301	.0294
1.9	.0287	.0281	.0274	.0268	.0262	.0256	.0250	.0244	.0239	.0233
2	.0228	.0222	.0217	.0212	.0207	.0202	.0197	.0192	.0188	.0183
2.1	.0179	.0174	.0170	.0166	.0162	.0158	.0154	.0150	.0146	.0143
2.2	.0139	.0136	.0132	.0129	.0125	.0122	.0119	.0116	.0113	.0110
2.3	.0107	.0104	.0102	.0099	.0096	.0094	.0091	.0089	.0087	.0084
2.4	.0082	.0080	.0078	.0075	.0073	.0071	.0069	.0068	.0066	.0064
2.5	.0062	.0060	.0059	.0057	.0055	.0054	.0052	.0051	.0049	.0048
2.6	.0047	.0045	.0044	.0043	.0041	.0040	.0039	.0038	.0037	.0036
2.7	.0035	.0034	.0033	.0032	.0031	.0030	.0029	.0028	.0027	.0026
2.8	.0026	.0025	.0024	.0023	.0023	.0022	.0021	.0021	.0020	.0019
2.9	.0019	.0018	.0018	.0017	.0016	.0016	.0015	.0015	.0014	.0014
3	.0013	.0013	.0013	.0012	.0012	.0011	.0011	.0011	.0010	.0010

Appendix H
Selected Full and Fractional Factorial Designs

2^2 Full Factorial

	A	B
1	–	–
2	+	–
3	–	+
4	+	+

All terms are free from aliasing.
Full Resolution

2^3 Half Fractional Factorial

	A	B	C
1	–	–	+
2	+	–	–
3	–	+	–
4	+	+	+

I + ABC
A + BC
B + AC
C + AB
Resolution III

2^3 Full Factorial

	A	B	C
1	−1	−1	−
2	1	−1	−
3	−1	1	−
4	1	1	−
5	−1	−1	+
6	1	−1	+
7	−1	1	+
8	1	1	+

All terms are free from aliasing.
Full Resolution

2^4 Half Fractional Factorial

	A	B	C	D
1	−	−	−	−
2	+	−	−	+
3	−	+	−	+
4	+	+	−	−
5	−	−	+	+
6	+	−	+	−
7	−	+	+	−
8	+	+	+	+

I + ABCD
A + BCD
B + ACD
C + ABD
D + ABC
AB + CD
AC + BD
AD + BC
Resolution IV

2^4 Full Factorial

	A	B	C	D
1	–	–	–	–
2	+	–	–	–
3	–	+	–	–
4	+	+	–	–
5	–	–	+	–
6	+	–	+	–
7	–	+	+	–
8	+	+	+	–
9	–	–	–	+
10	+	–	–	+
11	–	+	–	+
12	+	+	–	+
13	–	–	+	+
14	+	–	+	+
15	–	+	+	+
16	+	+	+	+

2^5 1/4 Fractional Factorial

	A	B	C	D	E
1	–	–	–	+	+
2	+	–	–	–	–
3	–	+	–	–	+
4	+	+	–	+	–
5	–	–	+	+	–
6	+	–	+	–	+
7	–	+	+	–	–
8	+	+	+	+	+

I + ABD + ACE + BCDE
A + BD + CE + ABCDE
B + AD + CDE + ABCE
C + AE + BDE + ABCD
D + AB + BCE + ACDE
E + AC + BCD + ABDE
BC + DE + ABE + ACD
BE + CD + ABC + ADE
Resolution III

2^5 Half Fractional Factorial

	A	B	C	D	E
1	–	–	–	–	+
2	+	–	–	–	–
3	–	+	–	–	–
4	+	+	–	–	+
5	–	–	+	–	–
6	+	–	+	–	+
7	–	+	+	–	+
8	+	+	+	–	–
9	–	–	–	+	–
10	+	–	–	+	+
11	–	+	–	+	+
12	+	+	–	+	–
13	–	–	+	+	+
14	+	–	+	+	–
15	–	+	+	+	–
16	+	+	+	+	+

I + ABCDE
A + BCDE
B + ACDE
C + ABDE
D + ABCE
E + ABCD
AB + CDE
AC + BDE
AD + BCE
AE + BCD
BC + ADE
BD + ACE
BE + ACD
CD + ABE
CE + ABD
DE + ABC
Resolution V

Appendix I

Selected Plackett-Burman Screening Designs

Plackett-Burman designs are Resolution III experiments; each main effect is partially confounded with all two-factor interactions not involving itself. These designs are intended to be used as screening designs to identify main effects, and are not meant to identify interactions.

Eight-Run 2^7 Plackett-Burman

	A	B	C	D	E	F	G
1	+	−	−	+	−	+	+
2	+	+	−	−	+	−	+
3	+	+	+	−	−	+	−
4	−	+	+	+	−	−	+
5	+	−	+	+	+	−	−
6	−	+	−	+	+	+	−
7	−	−	+	−	+	+	+
8	−	−	−	−	−	−	−

12-Run 2^{11} Plackett-Burman

	A	B	C	D	E	F	G	H	I	J	K
1	+	−	+	−	−	−	+	+	+	−	+
2	+	+	−	+	−	−	−	+	+	+	−
3	−	+	+	−	+	−	−	−	+	+	+
4	+	−	+	+	−	+	−	−	−	+	+
5	+	+	−	+	+	−	+	−	−	−	+
6	+	+	+	−	+	+	−	+	−	−	−
7	−	+	+	+	−	+	+	−	+	−	−
8	−	−	+	+	+	−	+	+	−	+	−
9	−	−	−	+	+	+	−	+	+	−	+
10	+	−	−	−	+	+	+	−	+	+	−
11	−	+	−	−	−	+	+	+	−	+	+
12	−	−	−	−	−	−	−	−	−	−	−

16-Run 2^{15} Plackett-Burman

	A	B	C	D	E	F	G	H	I	J	K	L	M	N	O
1	+	–	–	–	+	–	–	+	+	–	+	–	+	+	+
2	+	+	–	–	–	+	–	–	+	+	–	+	–	+	+
3	+	+	+	–	–	–	+	–	–	+	+	–	+	–	+
4	+	+	+	+	–	–	–	+	–	–	+	+	–	+	–
5	–	+	+	+	+	–	–	–	+	–	–	+	+	–	+
6	+	–	+	+	+	+	–	–	–	+	–	–	+	+	–
7	–	+	–	+	+	+	+	–	–	–	+	–	–	+	+
8	+	–	+	–	+	+	+	+	–	–	–	+	–	–	+
9	+	+	–	+	–	+	+	+	+	–	–	–	+	–	–
10	–	+	+	–	+	–	+	+	+	+	–	–	–	+	–
11	–	–	+	+	–	+	–	+	+	+	+	–	–	–	+
12	+	–	–	+	+	–	+	–	+	+	+	+	–	–	–
13	–	+	–	–	+	+	–	+	–	+	+	+	+	–	–
14	–	–	+	–	–	+	+	–	+	–	+	+	+	+	–
15	–	–	–	+	–	–	+	+	–	+	–	+	+	+	+
16	–	–	–	–	–	–	–	–	–	–	–	–	–	–	–

Appendix J
Selected Taguchi Designs

Taguchi L4 (2^3)

	1	**2**	**3**
1	1	1	1
2	1	2	2
3	2	1	2
4	2	2	1

	1	**2**	**3**
1		3	2
2			1

Taguchi L8 (2^7)

	1	**2**	**3**	**4**	**5**	**6**	**7**
1	1	1	1	1	1	1	1
2	1	1	1	2	2	2	2
3	1	2	2	1	1	2	2
4	1	2	2	2	2	1	1
5	2	1	2	1	2	1	2
6	2	1	2	2	1	2	1
7	2	2	1	1	2	2	1
8	2	2	1	2	1	1	2

	1	**2**	**3**	**4**	**5**	**6**	**7**
1		3	2	5	4	7	6
2			1	6	7	4	5
3				7	6	5	4
4					1	2	3
5						3	2
6							1

Taguchi L8 (1^4 with up to 2^4)

	1	**2**	**3**	**4**	**5**
1	1	1	1	1	1
2	1	2	2	2	2
3	2	1	1	2	2
4	2	2	2	1	1
5	3	1	2	1	2
6	3	2	1	2	1
7	4	1	2	2	1
8	4	2	1	1	2

The L8 (1^4 with up to 2^4) array does not have an interaction table.

Taguchi L9 (3^4)

	1	**2**	**3**	**4**
1	1	1	1	1
2	1	2	2	2
3	1	3	3	3
4	2	1	2	3
5	2	2	3	1
6	2	3	1	2
7	3	1	3	2
8	3	2	1	3
9	3	3	2	1

The L9 (3^4) array does not have an interaction table.

Taguchi L12 (2^{11})

	1	**2**	**3**	**4**	**5**	**6**	**7**	**8**	**9**	**10**	**11**
1	1	1	1	1	1	1	1	1	1	1	1
2	1	1	1	1	1	2	2	2	2	2	2
3	1	1	2	2	2	1	1	1	2	2	2
4	1	2	1	2	2	1	2	2	1	1	2
5	1	2	2	1	2	2	1	2	1	2	1
6	1	2	2	2	1	2	2	1	2	1	1
7	2	1	2	2	1	1	2	2	1	2	1
8	2	1	2	1	2	2	2	1	1	1	2
9	2	1	1	2	2	2	1	2	2	1	1
10	2	2	2	1	1	1	1	2	2	1	2
11	2	2	1	2	1	2	1	1	1	2	2
12	2	2	1	1	2	1	2	1	2	2	1

The L12 (2^{11}) array does not have an interaction table. This design is used to investigate main effects. The interactions are roughly distributed to all columns and cannot be used to analyze interactions.

Taguchi L16 (2^{15})

	1	**2**	**3**	**4**	**5**	**6**	**7**	**8**	**9**	**10**	**11**	**12**	**13**	**14**	**15**
1	1	1	1	1	1	1	1	1	1	1	1	1	1	1	1
2	1	1	1	1	1	1	1	2	2	2	2	2	2	2	2
3	1	1	1	2	2	2	2	1	1	1	1	2	2	2	2
4	1	1	1	2	2	2	2	2	2	2	2	1	1	1	1
5	1	2	2	1	1	2	2	1	1	2	2	1	1	2	2
6	1	2	2	1	1	2	2	2	2	1	1	2	2	1	1
7	1	2	2	2	2	1	1	1	1	2	2	2	2	1	1
8	1	2	2	2	2	1	1	2	2	1	1	1	1	2	2
9	2	1	2	1	2	1	2	1	2	1	2	1	2	1	2
10	2	1	2	1	2	1	2	2	1	2	1	2	1	2	1
11	2	1	2	2	1	2	1	1	2	1	2	2	1	2	1
12	2	1	2	2	1	2	1	2	1	2	1	1	2	1	2
13	2	2	1	1	2	2	1	1	2	2	1	1	2	2	1
14	2	2	1	1	2	2	1	2	1	1	2	2	1	1	2
15	2	2	1	2	1	1	2	1	2	2	1	2	1	1	2
16	2	2	1	2	1	1	2	2	1	1	2	1	2	2	1

	1	**2**	**3**	**4**	**5**	**6**	**7**	**8**	**9**	**10**	**11**	**12**	**13**	**14**	**15**
1		3	2	5	4	7	6	9	8	11	10	13	12	15	14
2			1	6	7	4	5	10	11	8	9	14	15	12	13
3				7	6	5	4	11	10	9	8	15	14	13	12
4					1	2	3	12	13	14	15	8	9	10	11
5						3	2	13	12	15	14	9	8	11	10
6							1	14	15	12	13	10	11	8	9
7								15	14	13	12	11	10	9	8
8									1	2	3	4	5	6	7
9										3	2	5	4	7	6
10											1	6	7	4	5
11												7	6	5	4
12													1	2	3
13														3	2
14															1

Appendix K
Selected Mixture Designs

Two-Factor Simplex Design

Run	A	B
1	1	0
2	0	1

Three-Factor Simplex Design

Run	A	B	C
1	1	0	0
2	0	1	0
3	0	0	1
4	0.5	0.5	0
5	0.5	0	0.5
6	0	0.5	0.5

Two-Factor Simplex Centroid Design

Run	A	B
1	1	0
2	0	1
3	0.5	0.5

Three-Factor Simplex Centroid Design

Run	A	B	C
1	1	0	0
2	0	1	0
3	0	0	1
4	0.5	0.5	0
5	0.5	0	0.5
6	0	0.5	0.5
7	0.333	0.333	0.333

Two-Factor Simplex Lattice Design

Run	A	B
1	1	0
2	0	1
3	0.5	0.5
4	0.75	0.25
5	0.25	0.75

Three-Factor Simplex Lattice Design

Run	A	B	C
1	1	0	0
2	0	1	0
3	0	0	1
4	0.333	0.333	0.333
5	0.667	0.167	0.167
6	0.167	0.667	0.167
7	0.167	0.167	0.667

Two-Factor Extreme Vertices Design

Run	A	B
1	0	1
2	1	0
3	0.5	0.5
4	0.25	0.75
5	0.75	0.25

Three-Factor Extreme Vertices Design

Run	A	B	C
1	0	0	1
2	1	0	0
3	0	1	0
4	0.33	0.33	0.33
5	0.17	0.17	0.67
6	0.67	0.17	0.17
7	0.17	0.67	0.17

Glossary

aliases—Effects that are identical except for the sign, that is, confounded effects.

alpha (α) risk—The probability of erroneously claiming a difference in two averages or two variances.

analysis matrix—The table that displays the design columns and any interaction columns with the outline of the calculations to be completed.

analysis of variance (ANOVA)—Basic statistical technique for analyzing experimental data. It subdivides the total variation of a data set into meaningful component parts associated with specific sources of variation, including interactions.

antagonism—An interaction that produces a response that is less than an individual factor.

attribute data—Characteristic or property that is measured in terms of pass/fail or go/no-go.

balanced design—Fractional factorial experiment design in which an equal number of trials is performed for each factor, at each combination of factor levels.

blocking—In a designed experiment, the process used to eliminate the effect of inhomogeneity in experimental material.

boldness—Term used to describe the choosing of the magnitude of the variable levels used within a response surface design. This concept suggests that the magnitude of variables should be large enough to capture the minimum and maximum responses under examination.

Box-Behnken design—An alternative to central composite designs when the optimum response is not located at the extremes of the experimental region or does not use previous results from a fractional design. All design variables must vary continuously.

calibration—Comparison of two instruments or measuring devices, one of which is a standard of known accuracy traceable to national standards, to detect, correlate, report, or to eliminate by adjustment any inaccuracy of the instrument or measuring device, as compared to the standard.

center points—With quantitative factors, the run defined when the levels of each factor are set at the midpoint of the high and low levels.

central composite—Finds the optimal levels of the design variables by adding a few more experiments to a full factorial design. All design variables must vary continuously.

characteristic—Factors, elements, or measures that define and differentiate a process, function, product, service, or other entity. A property that helps differentiate between items of a given sample or population. May be either quantitative (by variables) or qualitative (by attributes).

confounding—The result of defining columns for main effects and/or interactions that are identical or exactly reversed in sign. The resulting confounded effects are aliases of one another.

constrained design—A lower bound and/or an upper bound placed on some or all of the factors.

consumer's risk—In sampling, the potential risk that bad products will be accepted and shipped to the consumer.

control factors—See **factors**.

degrees of freedom (df)—Parameter that, in general, is the number of independent comparisons available to estimate a specific parameter and serves as a means of entering certain statistical tables. The number of unconstrained parameters in a statistical determination.

design matrix—The table that shows the settings for each factor for each run.

design of experiments (DOE)—A family of efficient techniques for studying more than one factor in a single experiment.

design points—The process conditions at which the response is measured.

destructive testing—Measurement, testing, and inspection of product or material that damages or destroys the product or materials so they are not usable. Contrast with nondestructive testing.

discrimination—Discrimination, or *resolution*, of a measurement system is its capability to detect and indicate even small changes in the measured characteristic. The main consideration when selecting or analyzing a measurement system.

d-optimal—Design variables that present multilinear constraints.

evolutionary operation (EVOP)—This technique adjusts variables in a process in small increments, in a planned pattern of changes, in the search for a more optimal point on the response scale. Although this approach may be slower than a DOE, the changes are small enough to prevent nonconformances, but large enough to establish changes. This process is accomplished in a production environment.

experiment—Process undertaken, usually empirically, to determine something that is not already known.

experimental design—Formal plan that details the specifics for conducting an experiment, such as which responses, factors, levels, blocks, treatments, and tools are to be used.

experimental error—The random noise in the experiment; an estimate of the inherent variation of the process.

extrapolate—Usually used in the context of using statistics to obtain an estimate or to infer a value beyond the known range; an educated guess based on known values or past practice.

factor analysis—Statistical technique that examines the relationships between a single dependent variable and multiple independent variables.

factorial experiment—A designed experiment where all possible treatment combinations are formed from two or more factors, each being studied at two or more versions (levels).

factors—Controllable (independent) variables that may cause a change in the response (dependent) variable.

foldover—The extension of a basic screening design by changing all signs; this assures that main effects will no longer be confounded with two-factor interactions in a combined analysis. Also called a *reflection*.

fractional factorials—Designs that contain a fraction of the number of full factorial runs. The resulting confounding is complete instead of partial. Can be effectively optimized for any number of factors.

***F*-test**—A test using the ratios of variances for pairs of samples. Used to determine whether the populations from which two samples were taken have the same standard deviation. The *F*-distribution is usually expressed as a table of the upper limit below which *F* can be expected to lie with some confidence level for samples of a specified number of degrees of freedom.

full factorial design—Designs that evaluate the combination of all levels of all factors in an experiment.

geometric screening design—Screening designs that are a power of two (for example, runs of 8, 16, 32, 64). The confounding of effects is complete. All fractional factorials are part of this family, as are Plackett-Burman designs for those run sizes.

Graeco-Latin squares—Extension of the basic Latin square designs where the number of blocking factors is greater than two.

heredity—The characteristic that factors with large effects will often have a significant interaction.

hierarchy of models—The preferred practice of including all main-effect terms in a model when their interaction is significant, regardless of the significance of the main effect terms.

hyper Graeco-Latin squares—Extension of the basic Latin square designs where the number of blocking factors is greater than two.

interaction—When the effect of factor A differs according to what the level factor of B is. A combined effect that cannot be predicted by the individual main effects.

Latin square design—Latin squares are a special form of fractional factorial design. They can be used as a form of blocking when there are two blocking factors to be used.

levels—The specific settings or options for a factor in an experiment.

line clearance—A procedure used to prevent mix-ups and/or comingling involving product, packaging, and labeling.

main effect—Average change in the response by changing the level of a factor averaged over all (levels) of other factors in a designed experiment. The main effect is the controlling factor; interaction effects may provide an indirect effect.

mean square—Sum of squares divided by degrees of freedom.

measurement systems analysis (MSA)—An analysis that considers operations, procedures, devices, and other equipment or personnel used to assign a value to the characteristic being measured.

mixture design—A special type of experiment for analyzing mixtures of components that must sum to a constant. This type of design is typically used in manufacturing of chemicals and food production. Triangular (ternary) graphs are used to display the mixture proportions.

model—An equation that is fit to the data.

noise—Variables that are difficult or impossible to control that affect the process. These are often environmental, such as ambient temperature and humidity.

nongeometric screening designs—Screening designs that are a multiple of four and not a power of two (for example, 12, 20, 24, 28, 36). The confounding of effects is partial. These designs are uniquely Plackett-Burman.

one-factor-at-a-time (OFAT) experiment—Each factor is varied individually while holding the levels of other factors constant. For example, an individual tries to fix a problem by making a change, then executing a test. Depending on the findings, something else may need to be tried.

outliers—An observed point that is distant from other observations; abnormal responses resulting from special causes or uncontrolled influences that occur during an experiment.

Pareto chart—A graphical tool for prioritizing effects.

Plackett-Burman designs—A family of screening designs that can address $n - 1$ factors with n trials so long as n is divisible by four. Designs can be geometric (the number of runs is a power of two) or nongeometric.

producer's risk—For a given sampling plan, refers to the probability of not accepting a lot, the quality of which has a designated numerical value representing a generally desirable level. Usually, the designated value will be the acceptable risk and quality level (AQL).

random effects—Factorial experiment where the variance of factors is investigated (as opposed to a fixed-effects model).

randomizing—Process used in statistics to avoid possible bias due to the influence of systematic disturbances that are either known or unknown.

refining designs—Designs that optimize a small number of factors, generally using more than two levels.

reflection—The extension of a basic screening design by changing all signs; this assures that main effects will no longer be confounded with two-factor interactions in a combined analysis. Also called a *foldover.*

repeat—A repetition of an experimental run without reapplying the setup for that run. It is not an independent repetition of the run.

replication—A repetition of a run that is a complete and independent reapplication of the setup.

residual—In a designed experiment, the differences between experimental responses and predicted values that are determined from a model.

residual error—In a designed experiment, the difference between the observed result and the predicted value (estimated treatment response) for that result, based on the empirically determined assumed model.

resolution—The degree to which estimated main effects are aliased (or confounded) with estimated two-level interactions, three-level interactions, and so on.

resolution I experiment—An experiment of exactly one run that only tests one level of a factor and cannot distinguish between the high and low levels of that factor.

resolution II experiment—An experiment in which main effects are confounded with other main effects.

resolution III experiment—An experiment in which two-factor interactions are aliased with main effects.

resolution IV experiment—An experiment in which two-factor interactions are aliased with other two-factor interactions, but not with main effects. Main effects are aliased with three-factor interactions.

resolution V experiment—An experiment in which two-factor interactions are not aliased with each other, but are aliased with three-factor interactions.

resolution VI experiment—An experiment in which estimates of main factors and two-factor effects are not confounded with each other or with three-factor interactions. Three-factor and higher interactions may be confounded with one another.

resolution VII experiment—An experiment in which estimates of main factors, two-factor effects, and three-factor effects are not confounded with one another but may be confounded with higher-order interactions. Four-factor and higher interactions may be confounded.

response—The output from the experiment; the dependent variable(s).

robust design—Design capable of properly performing its intended functions under a wide range of conditions, including some anticipated level of customer or user abuse.

robustness—A condition in which changes in the setting of a factor do not have an impact on the response variable; that is, the response is not sensitive to noise in the factor settings.

run—A defined combination of the levels for all factors. Also called a *treatment.*

screening designs—Designs that provide information primarily on main effects, incurring the risks associated with confounding, in order to determine which of a large number of factors merit further study.

screening experiment—First step of a multiple factorial experiment strategy, where the experiment primarily assesses the significance of main effects. Two-factor interactions are normally considered in the experiments that follow a screening experiment.

signal factors—Factors that can be set or controlled by the operator of the process.

signal-to-noise ratio (S/N)—A measure of the impact of noise factors on performance. The larger the S/N, the more robust the product is against noise.

simplex axial designs—A design that consists mainly of the points positioned inside the simplex.

simplex centroid design—A design that only includes the centroid points. Components that appear in a run in a simplex centroid design have the same values.

simplex lattice design—A space-filling design that creates a triangular grid of runs. The design is the set of all combinations where the sum of the factors is 1.

sparsity—The characteristic that main effects are generally larger than two-factor interactions, while three-factor interactions can usually be ignored.

Student's *t*-test—A method for assessing whether the means of two groups are statistically different from each other.

sum of squares—Summation of the squared deviations relative to zero, to two level means, or the grand mean of an experiment.

synergism—An interaction that produces a response that is greater than an individual factor.

test method validation (TMV)—See **measurement systems analysis**.

treatment—A defined combination of the levels for all factors. Also called a *run*.

type I error—An incorrect decision to reject something (such as a statistical hypothesis or a lot of products) when it is acceptable.

type II error—An incorrect decision to accept something (such as a statistical hypothesis or a lot of products) when it is unacceptable.

variables data—Measurement quantities that are measured on a continuous and infinite scale.

variance—The square of the standard deviation.

variation—A change in data, characteristic, or function caused by one of four factors: special causes, common causes, tampering, or structural variation. Sources of variation include time to time, part to part, and within the sample.

Bibliography

Anderson, Mark J., and Patrick J. Whitcomb. 2005. *RSM Simplified: Optimizing Processes Using Response Surface Methods for Design of Experiments.* New York: CRC Press.

———. 2007. *DOE Simplified: Practical Tools for Effective Experimentation.* 2nd ed. New York: Productivity Press.

Barrentine, Larry B. 1999. *An Introduction to Design of Experiments: A Simplified Approach.* Milwaukee: ASQ Quality Press.

Box, George E. P., J. Stuart Hunter, and William G. Hunter. 2005. *Statistics for Experimenters: Design, Innovation, and Discovery.* 2nd ed. New York: John Wiley & Sons.

Box, George E. P., William G. Hunter, and J. Stuart Hunter. 1978. *Statistics for Experimenters: An Introduction to Design, Data Analysis, and Model Building.* New York: John Wiley & Sons.

Cornell, John A. 1990. *How to Run Mixture Experiments for Product Quality.* Vol. 5 of *The ASQ Basic References in Quality Control: Statistical Techniques.* Milwaukee: ASQ Quality Press.

Durivage, Mark Allen. 2014. *Practical Engineering, Process, and Reliability Statistics.* Milwaukee: ASQ Quality Press.

———. 2015. *Practical Attribute and Variable Measurement Systems Analysis (MSA): A Guide for Conducting Gage R&R Studies and Test Method Validations.* Milwaukee: ASQ Quality Press.

Ful-Chiang Wu and Chi-Hao Yeh. 2006. "A Comparative Study on Optimization Methods for Experiments with Ordered Categorical Data." *Computers & Industrial Engineering* 50 (3): 220–32.

Iglewicz, Boris, and David C. Hoaglin. 1993. *How to Detect and Handle Outliers.* Vol 16. Milwaukee: ASQ Quality Press.

Mehta, Bavan "Bob." 2013. *Implementing ISO/IEC 17024:2005: A Practical Guide.* Milwaukee: ASQ Quality Press.

Montgomery, Douglas C. 2013. *Design and Analysis of Experiments.* 8th ed. New York: John Wiley & Sons.

Ross, Phillip J. 1996. *Taguchi Techniques for Quality Engineering.* 2nd ed. New York: McGraw-Hill.

Roy, Ranjit K. 1990. *A Primer on the Taguchi Method.* Dearborn, MI: SME.

Roy, Ranjit K. 2001. *Design of Experiments Using the Taguchi Approach: 16 Steps to Product and Process Improvement.* New York: John Wiley & Sons.

Schmidt, Stephen R., and Robert G. Launsby. 1992. *Understanding Industrial Designed Experiments.* 3rd ed. Colorado Springs: Air Academy Press.

Siebels, Don L. 2004. *The Quality Improvement Glossary.* Milwaukee: ASQ Quality Press.

Walfish, Steven. 2006. "A Review of Statistical Outlier Methods." *Pharmaceutical Technology* (November).

Wheeler, Donald J. 1990. *Understanding Industrial Experimentation*. 2nd ed. Knoxville, TN: SPC Press.

Zimmerman, Steven M., and Marjorie L. Icenogle. 2003. *Statistical Quality Control Using Excel*. 2nd ed. Milwaukee: ASQ Quality Press.

WEBSITES

http://asq.org/glossary/
http://en.wikipedia.org/wiki/Analysis_of_variance
http://en.wikipedia.org/wiki/Fractional_factorial_design
http://en.wikipedia.org/wiki/One-way_ANOVA
http://en.wikipedia.org/wiki/Two-way_analysis_of_variance

Index

F

G

H

I

L

M

N

O

P

R

S

T

U

V

W

X

Z

NOTES

NOTES

www.ingramcontent.com/pod-product-compliance
Lightning Source LLC
Chambersburg PA
CBHW081213180926
51613CB00003B/223
* 9 7 8 0 8 7 3 8 9 9 2 4 6 *